This book is a condensation of Jere:
Free: The End of The Huma
which was published in Au~~s~~__

Jeremy Griffith was born in Australia in 1945, raised on a sheep station in central New South Wales and educated at Geelong Grammar School in Victoria. After graduating from Sydney University with a degree in biology he undertook the most thorough investigation yet carried out into the plight of the now-believed-extinct Tasmanian Tiger. During the six year period he spent in the wilds of Tasmania Jeremy's search and findings received international scientific and popular media coverage.

It was during this time in Tasmania that he turned his attention to the plight of another species — humanity. He says that *Free: The End of The Human Condition,* which he went on to spend 13 years writing, 'grew out of my desperate need to reconcile my extreme idealism with reality'.

After leaving Tasmania Jeremy established a successful furniture manufacturing business with one of his brothers. He recently disposed of his interest in the company.

Jeremy has given many lectures on his conclusions and welcomes most opportunities to explain them.

Tim Macartney-Snape was born in Tanzania in 1956 and came to Australia at the age of 12. One of the world's leading mountaineers, he was one of the first Australians to climb Mt Everest, for which he was awarded the Order of Australia medal. In 1990 he climbed Mt Everest a second time, again without supplementary oxygen but this time unassisted and unroped, and by first walking 1,000 kilometres from sea level to base camp with his wife Ann and sister Pip. The film and book about this remarkable journey will appear in 1992.

Tim was educated at Geelong Grammar School and graduated in environmental science from the Australian National University. Formerly a director of the trekking company Wilderness Expeditions, he is writing a book to encourage others to venture into the natural world safely and responsibly.

Tim is in constant demand to speak about his mountaineering experiences and his source of motivation in this most demanding of human endeavours. The information in *Free . . .* has helped him immensely to understand his motivation and he promotes Jeremy's insights enthusiastically and lectures on them.

BEYOND THE HUMAN CONDITION

Jeremy Griffith

with a Foreword by
Tim Macartney-Snape

Published by
Foundation For Humanity's Adulthood
GPO Box 5095, Sydney NSW 2001, Australia
Phone: (02) 9486 3308
Website: **www.humancondition.info**

The crux problem on Earth has always been the human condition, our capacity for good and evil. The discovery of its cause is the key to our freedom. Humanity can now end its insecure, upset adolescence and enter its peaceful adulthood.

Special thanks to:
Annie Williams, Simon Griffith, Tim Macartney-Snape,
Rosy and Howard Whelan, Phil McSharry, Sally Kaufmann,
Steve van Hemert, Mike Rigg, Bahram Boutorabi,
Professor John Wren-Lewis, Christopher Stephen, James West,
Tim Watson and Bill Ryles.

Cover: *Cringing in Terror* (c 1794–96) and *Albion Arose* (c 1794–96)
by William Blake; coloured impression of *Albion Arose* by Carol
Marando. Design by Deborah Brash.

First published in 1991 by:
Foundation For Humanity's Adulthood
GPO Box 5095, Sydney NSW 2001, Australia
Phone: (02) 486 3308

Foreword © 1991 Tim Macartney-Snape.
Beyond The Human Condition and illustrations for
'The Development and Resolution of Upset' and 'Adam Stork'
© 1991 Jeremy Griffith. Charts © 1988 Jeremy Griffith.

NIV Bible verses by permission Hodder and Stoughton.

ISBN 0 646 03994 6
CIP — Biological Psychology

Typeset by Wallace & Schoenauer, Sydney.
Printed by Dai Nippon, Hong Kong.

This book is dedicated to the vision of Sir Laurens van der Post:

' . . . for I had a private hope of the utmost importance to me. The Bushman's physical shape combined those of a child and a man: I surmised that examination of his inner life might reveal a pattern which reconciled the spiritual opposites in the human being and made him whole . . . it might start the first movement towards a reconciliation . . . '
Laurens van der Post, *The Heart of the Hunter*, 1961.

And that of Dr Louis Leakey who believed:

' . . . that knowledge of the past would help us to understand and possibly control the future.'
Mentioned by Dr Mary Leakey in her book *Disclosing the Past*, 1984.

Contents

Tim Macartney-Snape holds the
Foundation For Humanity's Adulthood flag
on the summit of Mount Everest, May 11, 1990.

Foreword

by Tim Macartney-Snape

CHARLES DARWIN'S *The Origin of Species* (1859) was one of the great milestones in the heroic journey of human ideas. In retrospect, Darwin's theory was a very simple concept to grasp, even though it was not thought of until well after one of the giants of science, Isaac Newton, had invented calculus and worked out, almost perfectly, the laws of motion and the movement of the planets. While Newton's ideas helped us to begin understanding the physical world, we had to wait for Darwin's ideas on evolution to begin understanding how life on Earth developed into the complex array of organisms that exists today. With an understanding of the basic idea that evolution is a genetic learning process (nature's 'trial and error' method if you like, where organisms are refined and improved through their genes), that evolutionary change happens slowly and by using our subsequent knowledge to interpret the fossilised remains of prehistoric life forms, we have finally found ourselves in a position to begin understanding our own origins.

This is indeed a triumph for the mind. For all recorded history we have struggled to find meaning in our existence, inventing elaborate stories to explain our origins. Rational thinking and its product, science, have brought us a long way from when we recorded our myths on the walls of caves. Science has been a great winnower of superstition, but its mechanistic approach has had the effect of excluding us from the explanations that I believe we needed and set out to find in the first place.

Is our thirst for meaning without justification or is there some hidden meaning to life? Jeremy Griffith deals with this question in these pages and in his main work, *Free: The End of The Human Condition.*

Jeremy may well be ahead of his time because his explanations lead to a great shift in the way we think. As for our need for his ideas, they may have arrived only in the nick of time. A handful of physicists, mathematicians and a few experimental scientists are beginning to hint that evidence suggests a purpose in the development of life. This is very bold indeed for as you will see as you read on, there are real problems with the scientific mind accepting this notion.

The science of life, biology, will generally deny any teleological or purposeful explanations for life and although many biologists hold religious beliefs, there has been no true, sound, scientific effort to solve the mystery of the mind's obsession for meaning. There has been little effort to interpret our moral codes. In fact despite the vast inroads made by the mind into the uncharted territory of the universe, we still know very little about ourselves, about our psychological development and the functioning of our 'on-board necktop computer'.

The human mind is capable of unsurpassed sensitivity to our planet. This becomes abundantly clear to me when I react to things I find beautiful in nature such as a flower or a mountain landscape. Despite suffering from lack of oxygen it was vividly clear to me when, after struggling up the north face of Mt Everest in 1984, I arrived at the summit to view a world bathed in the rosy glow of the setting sun and felt an almost incandescent warmth and love towards our planet. This sensitivity is clearly shown in our expression of wonder at natural beauty reflected in our art, music and poetry and is probably clearest of all when examining unconditional love. This emotion is so intense it seems to transcend explanation, making it even more inaccessible to scientific scrutiny.

Despite our great sensitivity and propensity for love and perception of goodness, there is also a dark side to our history: an equal propensity for evil. We are undoubtedly the most ferocious and destructive force and the cruellest animal that has ever lived on this planet. To say that we have outgrown this by the process of civilisation is in my view, untrue. One only needs to look at the recent wars in the Persian Gulf to see that 10,000 years of civilisation have done little to curb our anger, aggression and cruelty. That region is the cradle of civilisation yet from an overview of our behaviour, all that has changed is the means by which we express our anger. It is true I think that the civilising process has taught us control but it seems to have done nothing to alter our nature. The person who appreciates Michelangelo's art is still capable of murder.

This duality of good and evil, the essence of the human condition, has no doubt perplexed humanity from the dawn of the process we call thinking. It is only now, by virtue of what we have learnt about the mechanisms of evolution and human prehistory, that we can at last hope to reach an understanding. I believe the reader will discover in these pages as I have, that Jeremy Griffith has indeed achieved this understanding. Perhaps even more astonishing is that he brings us the realisation that the duality of human nature, the good and evil, is part of an essential process in the long journey to this remarkable understanding.

As the mind used scientific enquiry to develop an understanding of our world, we put it successfully to work in the cause of goodness by stamping out untold misery and suffering in humanity. Diseases were cured, life made longer and healthier and transport and communications were

11

miraculously improved. Science has brought great comfort to millions of people. Similarly, it has also been harnessed to commit new and hitherto undreamt of miseries in the cause of evil. Despite having come so far, the question of the origin of our 'evil' side remains more urgent than ever. Will we, as a species, always have the propensity to create atrocities such as the holocaust and the bombing of Hiroshima, or is this merely a phase in human development which has to be overcome in the process of our evolution?

From this fundamental question a host of other increasingly urgent questions arise which must concern anyone with a broad view of the human situation.

- Why, if we are as sensitive as our emotions and art indicate, do we seem not to care that we are out of balance with the planet we 'grew up' on?
- Why do we create so much ugliness and surround ourselves with vulgarity that is so removed from what is classically accepted as beautiful?
- Why can't we halt the alarming change we are making to our life-giving atmosphere by the profligate burning of fossil fuels, the destruction of tropical forests and the depletion of the ozone layer?
- Why can't we control our burgeoning population that will surely, unless we act immediately, lead to our destruction?
- Why do we gear our economic systems and consumption patterns to growth when we live in a world where the limits to growth are obvious and are speeding up to confront us head-on with an almighty crash?

If we bring war and the countless other afflictions caused by human greed and ambition into the picture we very quickly have to start putting up our thought barriers. Without these barriers we could not cope in the ever increasing pace of modern life. One of the great afflictions of our time is our inability to see the reality of our predicament, yet it is that very ability to cling to the positive that has kept us afloat in a sea of doubt. But the time has come when we must confront reality and face the real problem, the cause of our capacity for evil.

Certainly many noble efforts are made by many people to try to redress the balance. The environment and human rights movements are examples. Surely though, trying to save whales trapped in the Arctic ice and to stop the ozone hole growing is merely plugging the valves when the real problem is the pressure of our upset nature. We have been grappling with the symptoms of our often destructive, insensitive, egotistical and aggressive nature rather than with its psychological cause. The fundamental problem on earth is our upset state or condition, not the expressions of it. The greatest challenge we face is not to escape into the solar system carrying with us the human condition but to use our minds to alleviate our condition, thus clearing the way for a saner, healthier, more stable future.

It was not until I studied biology at university that I fully realised that our genus *Homo* is the product of around 2 million years of evolution and that our primate ancestors were evolving for around 10 million years before that. In our most refined forms as *Homo sapiens* and *Homo sapiens sapiens*, we have been around for nearly half a million years. For more than 99 per cent of that time we lived as hunter-gatherers. It is only in the past 10,000 years, merely 0.5 per cent of *Homo's* existence, that we have led a settled lifestyle. Evolution is an imperceptibly slow process and those 10,000 years are but a few seconds on development's clock. Most of our refinement to what we are today was geared to perfecting us for life in the open, plenty of exercise, an 80 per cent vegetarian diet, and most important of all, living cooperatively in groups. Just by recognising those facts and trying to practise those dietary, physical and social patterns and trying to put them into a modern context we would immediately solve many of our current problems. Why isn't that taught in school?

The main question however is: why in the process of civilisation and the acquisition of knowledge did we lose our affinity with our natural surroundings and each other?

When renowned author Sir Laurens van der Post made a television documentary about the few remaining Bushmen of the Kalahari, the people most closely related to our hunter-gatherer forebears, world wide response was overwhelming. (This primitive, relatively innocent race once thrived all over southern Africa but was exterminated in all areas except the Kalahari desert, first by more sophisticated black races and later by even more sophisticated white races.) Many people, including myself, found the Bushmen endearing and could relate to their natural way of life, which we have lost in the civilising process. Their social structure of an extended, happy family and their spontaneous sense of fun and happiness were perhaps the most touching of all and exemplified qualities that we lack in our modern age.

Similarly, the simple folk who live in Nepal's lush and rugged foothills have appealed to me since I first visited that country. Unlike the Bushmen, these people are not at a stone age stage of development, but by virtue of their isolation and difficult living conditions, they exhibit a comparative innocence that we have lost. Most people I know who have been in contact with them react in the same way: somewhere deep inside a chord of wonder is struck and they cling to it as something very valuable.

The inevitable question is why are these qualities, so alive in the primitive and so deeply felt and revered by us, so suppressed in us today? Jeremy Griffith answers this question unequivocally in these pages. We have had to follow the path we have taken for a very good reason. An inevitable part of the great search for understanding was that we had to depart from an ideal state in order to gain a profound understanding of it. The search ends with Jeremy's revelations.

In their book *New Mind, New World* (1989), Paul Ehrlich and Robert Ornstein proposed that we fail to think globally because our evolutionary (i.e. hunter-gatherer) circumstances make us most capable of responding to our familiar and immediate surroundings and so we respond more positively to a child being stuck in a well than to the fact that thousands of children are dying from malnutrition every day. That may be partially true but it can't be the full truth. If it is, why is the human race as a whole so alienated from the wilderness it grew up in, why do we huddle in cities and deny the world that was our original home?

If you accept the concept that evolution always leads to a more successful 'model' for the prevailing conditions, then on a superficial level you might conclude that perhaps we are slowly becoming better. Here we should halt to briefly explore what we mean by 'better' and indeed, 'goodness'. Jeremy's convincing definition that I think most people would agree with is that, as a basic foundation for life, order is imperative and therefore order rather than chaos is the framework for a better, more meaningful life and that the ultimate state of goodness would be a state of universal love. These, to put it one way, are the consistent and universal social ideals: they have been determined separately by most of the great civilisations and religions in history.

So to get back to evolution; if we were slowly evolving towards a 'better' state, universal love and order rather than chaos would be the ultimate goal. Looking at the present evidence it quickly becomes clear that although we may appear to be becoming more civilised, or to put it another way, learning to subdue our badness or 'upset' as Jeremy calls it, we are merely bottling it up. Instead of being blatant, taking the form of mass genocide for instance, our upset is becoming manifest in more subtle ways — through the disintegration of family life, psychoses, drug abuse, crime, insensitivity to the environment, etc. We do not appear to be succeeding! No wonder we look at our fellow humans in the Kalahari and the Himalaya with a sense of longing.

Why, after knowing the perfect social ideals for thousands of years, have we not been able to achieve them? The answer is not in our physical evolution that has brought us so far, but in what went on in our heads when we first became conscious, and the ensuing evolution of our ideas. Those ideas incorporate much of what science has discovered about the process of the development of life. For very valid reasons that Jeremy explains clearly, we have evaded putting the evidence together and finding the fundamental understanding that we need. Remarkably, Jeremy has brought the evidence together and most importantly, he explains that the terrible part of humanity's journey has been an unavoidable and necessary evil.

To put it simply, Jeremy Griffith reveals that the human condition developed when our conscious mind emerged, some 2 million years ago. At that time, a battle began between the conscious mind and our instinctive self that could only be resolved through an understanding of the difference between consciousness and instinct. The conflict arose because the conscious mind is a memory-association-based learning system, which means it is insightful or understanding, whereas instincts are derived from the genetically-based learning system, which is not insightful.

Jeremy explains how, by the time the conscious mind emerged, we had already acquired an instinctive orientation to the 'ideal' of being cooperative or of creating greater order. Of course this was not a conscious understanding of that ideal. The battle began when the conscious mind embarked on the necessary voyage to find understanding of that ideal and everything else in life. Unavoidably and tragically, the instinctive self 'criticised' the search because it was not able to recognise and tolerate the misunderstandings and mistakes that the conscious mind made while trying to find reasons for everything. To find understanding, the conscious mind had to defy the ignorant criticism levelled at it by the instinctive self. Our egocentricity is the expression of our embattled conscious thinking selves.

16

This necessary battle to defy our instinctive self or soul and find understanding left us angry, egocentric and alienated. Being insecure, being unable to refute the criticism with explanation of why these mistakes were necessary, we coped by attacking the criticism, by trying to prove it wrong and by blocking it out or repressing it. Becoming upset was inevitable while we lacked the defence for our mistakes. We had to endure becoming upset if we were to understand why we had to make mistakes. It follows that having discovered why we had to make mistakes, our upset 'human nature' — our anger, egocentricity and alienation — can subside. Understanding what happened in our species' development — and the rehabilitation that automatically follows it — is the real or profound solution to the pressure on earth caused by our insecure, upset and destructive nature.

Finding that understanding took the combined effort of humanity 2 million years! It involved not only Charles Darwin with his revolutionary ideas, but people like Carnot, Kelvin and Clausius and the Laws of Thermodynamics. It involved Ilya Prigogine discovering the 'Second Path' of the Second Law of Thermodynamics (also termed 'negative entropy') in which systems of matter in disequilibrium integrate instead of disintegrating and so develop larger and more stable wholes or greater order. It involved Watson and Crick who made the fantastic breakthrough in determining the structure of DNA, the molecular carrier of genetic information. It involved the work of people like Drs Louis and Mary Leakey, pioneers in palaeoanthropology. In fact in one way or another it involved all of humanity.

Jeremy Griffith's concepts are without doubt confronting. I cannot caution the reader too strongly that there will be incredulity then discomfort when you begin to realise the depths to which this information takes us, but you must be strong because to go beyond the human condition requires us to go through those depths which have hitherto been forbidden territory. The total relief from our condition that awaits us on the other side makes the struggle well worth while, if not imperative.

When I think of the depth of uncharted territory in our mind I am reminded of a favourite quote of Sir Laurens van der Post (from Gerard Manley Hopkins): 'O the mind, mind has mountains; cliffs of fall Frightful, sheer, no-man fathomed.' Having at last been safely guided through this region of our repressed self or psyche most of us will be fearfully numbed at first, but the momentum of reason will carry us through this inevitable phase of the reconciling process.

At last we have the knowledge that will allow us to triumphantly climb from the dark depths of our searching on to the uplands of our ideals which it has been our destiny to reach.

The proof of any idea depends on its ability to explain the situation it applies to. For me, Jeremy's explanations have clarified so much that was inexplicable about myself and what goes on in the world. It is like having mist lift from country you've never seen in clear weather. Our age old reliance on faith and belief is over now that we know our direction. Our new-found understanding brings such relief and beauty that it makes the wonder we once bestowed upon mystery appear trivial. Even though we are unable to change ourselves immediately, we can participate in the new order right from the start by recognising the truth and supporting it without reproach or guilt.

In time there will be elaborations written and spoken about Jeremy Griffith's concepts. I believe that eventually, more words of importance will be spoken and written about them than about any others — ever. Eventually everyone will understand them, perhaps not for a few generations, but the sooner we all come to terms with them the richer and more exciting our lives will be and the sooner we will bring about a real repair of our planet.

Introduction

HERE on Earth, some of the most complex arrangements of matter in the known universe have come into existence. Life with its incredible diversity and richness developed.

By virtue of our mind, the human species must surely be the culmination of this grand experiment of nature that we call life. As far as we can detect, we are the first organism to have developed the ability to think and reflect upon itself. In this world, ravaged as it is by strife, it is easy to lose sight of the utter magnificence of what we are. The human mind must be nature's most astonishing creation.

One of the greatest demonstrations of our intellectual brilliance was sending three of our kind, in a machine of our own invention, to the Moon and back.

How far we have come!

But what a mess our world is in!

Despite the tremendous successes science has brought us, our plight only seems to be worsening in terms of human happiness and the Earth's well-being. Better forms of management such as better laws, better politics, better economics and better self-management such as new ways of disciplining, organising, or even transcending our upset natures have all failed to end our destructiveness and bring us peace and happiness. The recent war in the Persian Gulf, with its real threat of nuclear and chemical devastation, its scenes of torture, massacre and environmental destruction, has convinced many that nothing has changed for the better.

As we approach the twenty-first century man is starting to realise that the underlying problem is psychological. We are beginning to suspect that war, overpopulation, environmental degradation, resource depletion, species extinction, drugs, starving millions, crime, family breakdown, despair and even sickness are merely symptoms of a deeper problem: our often destructive, insensitive, egotistical and aggressive nature.

Environmental issues dominate our concern but surely we are focusing on the symptoms not the cause of the problem. The real issue is ourselves, our upset state or nature or condition. The real frontier, challenge and adventure before us now is not outer space, as we tend to believe, but inner space — the human condition no less.

> **. . . we need a new kind of explorer, a new kind of pathfinder, human beings who, now that the physical world is spread out before us like an open book . . . are ready to turn and explore in a new dimension.**
> Laurens van der Post, *The Dark Eye in Africa*, 1955.

The time has come — and it is now a matter of urgency — to switch our attention from the physical to the psychological and to grapple with the crux problem, that greatest of all paradoxes, the riddle of human nature. If the universally accepted ideals are to be cooperative, loving and selfless — they have been accepted by the great civilisations as the basis for their constitutions and laws and by the founders of all the great religions as the basis of their teachings — why are we competitive, aggressive and selfish? What is the reason for our divisive nature?

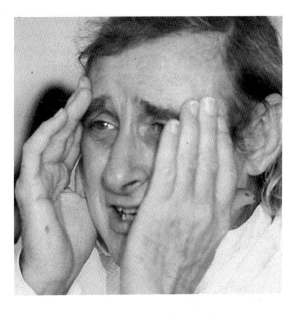

Comedian Spike Milligan, interviewed by John Newton.

Newton: Bob Ellis — the Australian playwright — once said 'humour is the sparkle on the deep river'.

Milligan: That's quite beautiful . . . the sparkle on the deep river, yes!

Newton: That being the case, can we use humour to clean up the deep river? Or must the poor clown, as you said earlier, stand by and watch as the human race rushes towards annihilation?

Milligan: Tell 'em jokes as they go by. That's the awful part of it. I don't think laughter can save us, not at all. It is a serious business. We can laugh during it, at it. The world has never reached this hour before. The politicians are totally unprophetic. They run the world politically, believing it to be the right thing, and it's turned out to be a total disaster. And they don't know what to do.

Text and Photo: *The Bulletin*, December 26, 1989.

The world is hurtling to catastrophe: from nuclear horrors, a wrecked ecosystem, 20 million dead each year from malnutrition, 600 million chronically hungry . . . All these crises are man made, their causes are psychological. The cures must come from this same source; which means the planet needs psychological maturity . . . fast. We are locked in a race between self destruction and self discovery.
Australian journalist and author Richard Neville, The *Sydney Morning Herald, Good Weekend* supplement, October 14, 1986.

The purpose of this book is to unlock the riddle of human nature; to answer this question of questions of the origin of, or reason for, our so-called evil side.

The explanation to be presented is confronting — that cannot be avoided — but it is also liberating. Paradoxically, and this is why it is liberating, the answer dignifies humans in the most remarkable way. It lifts the burden of guilt from humanity. It exonerates us and restores our love of ourselves, our kind and all creation. It allows us to heal the wounds we have inflicted on the planet, on one another and on ourselves. It provides us with the key to freedom from our upset state or condition. It brings us understanding of ourselves.

Human nature is not the immutable state we have often considered it to be; rather it is the symptom of a condition — the human condition — that will disappear now that the condition is relieved.

The eternal question has been why 'evil'? Why are we humans capable of extreme greed, hatred, brutality, rape, murder, war? Are we essentially good, and if so, what is the cause of our evil, destructive, insensitive and cruel side? Does our inconsistency with the universally accepted ideals of love, cooperation and selflessness mean we are essentially bad? Are we a flawed species, a mistake — or are we possibly divine beings? Life has always been a constant struggle to understand the conflicting forces within us.

Insecure in our goodness and aware of our badness, humanity has through the ages searched for a clear understanding of this great paradox. Neither philosophy nor science has been able to give a clarifying explanation. Religious assurances such as 'God loves you' don't explain why we are lovable. Lacking clear understanding, we still carry a burden of unhappiness within ourselves.

When the stars threw down their spears
And watered heaven with their tears,
Did he smile his work to see?
Did he who made the lamb make thee?
William Blake, *The Tyger, c* 1789–93.

If we call the embodiment of the ideals that sustain our society 'God' then we're a 'God fearing' species, condemned by the ideals, in fear and insecurity. Our human predicament, or condition, is that we have had to live with an undeserved sense of guilt.

Whenever we have tried to understand why there was evil in the world and indeed in ourselves we couldn't find an answer and eventually had to put the question out of our minds. We have coped with our sense of guilt by blocking it out, sensibly avoiding the whole depressing issue of our inconsistency with the ideals.

So skilled are we at overlooking the hypocrisy of human life and blocking out the question it raises of our guilt or otherwise, when the hypocrisy does appear we often fail to recognise it or the question it raises. In fact we couldn't afford to mention the human condition until it could be solved. While 'human nature' appears in dictionaries, 'human condition' doesn't.

Where adults now fail to recognise the paradox of human behaviour, children in their naivety still do. They ask, 'Mummy why do you and Daddy shout at each other?' and 'why are we going to a lavish party when that family down the road is poor?' and 'why do men kill one another?' and 'why is everyone unhappy and preoccupied?' and 'why are people so artificial and false?'.

The truth is the hypocrisy of human behaviour is all about us. Two-thirds of the people in the world are starving while the rest bathe in material security and go on seeking still more wealth and luxury.

> **With regard to the hypocrisy of life, in 1969 in the southern states of the USA there was a problem with busing — negroes weren't allowed on buses designated for whites — and I remember at that time someone saying: 'We can get a man on the moon, but we still can't get a negro on a bus!'**
> Bob Smith, a supporter of the Foundation For Humanity's Adulthood.

Humans can be heartbroken when they lose a loved one but are also capable of shooting one of their own family. We have dived into raging torrents to help others without thought of self but have also molested children. We have tortured one another but have also been so loving we regularly gave our lives for others. A community will pool their efforts to save a kitten stranded up a tree and yet humans will also **eat elaborately prepared dishes featuring endangered animals** (*Time* magazine, April 8, 1991). They have been sensitive enough to create the beauty of the Sistine Chapel, yet so insensitive as to pollute their planet to the point of threatening their own existence.

Good or bad, loving or hateful, angels or devils, constructive or destructive, sensitive or insensitive, what are we? Throughout our history, we've struggled to find meaning in the awesome contradictions of the human condition. We desperately need a clear biological understanding of ourselves, understanding that will liberate us from criticism, lift the burden of guilt, give us meaning, bring peace to our minds and lead us to achieving our psychological maturity as a species. Catch-phrases like 'self-esteem' and 'human potential' stress this yearning for self-justification and self-realisation. To adapt a famous expression of Benjamin Disraeli's, stalled halfway between ape and angel is no place to stop.

What then is the answer to this question of questions; this problem of good and evil that defies us, or is there no answer?

> **The problem of the origin and universality of sin . . . is probably one of those problems which the human mind can never satisfactorily answer.**
> *The Bible Reader's Encyclopedia and Concordance* (under 'sin').

> [In his book *Grist for the Mill* (1977), Ram Das, when asked] **why did we** [fall from grace] **in the first place?** [replied] **that is the question which is the ultimate question** [and that] **your subject-object mind can't know the answer to that question.**

Before Darwin, the origin of the variety of life on Earth seemed inexplicable. And yet his idea of natural selection was so simple that a contemporary, Thomas Henry Huxley, responded, **How extremely stupid of me not to have thought of that!**

While the crux question of our time of good and evil has always seemed inexplicable, in fact it too has an amazingly simple answer.

What follows is a journey that explores the biology of the human condition from a beginning that preceded conscious thought. It traces human development through to the twentieth century and looks at the source of our current distress. Though the ideas are in essence quite simple, their challenging nature can make them difficult to grasp. To aid understanding, analogies and illustrations have been used, which though simplistic, have already proved effective.

We begin with an analogy using migrating birds, which helps us to understand the essence of the human condition. Don't worry about whether storks eat apples or talk — that is irrelevant — the purpose of this little story is to consider what happened when two conflicting forms of thinking had to share the same brain.

The Story of Adam Stork

MANY bird species are perfectly orientated to instinctive migratory flight paths. Every winter, without ever 'learning' where to go and without knowing why, they fly to warmer feeding grounds and return in summer to breed.

Consider a flock of migrating storks returning from southern Africa to their summer breeding places on the roof-tops in Europe.

Suppose that in the instinct-controlled brain of one of them (let's call him Adam) we place a fully conscious mind. As he flies northwards, he sees an island off to the left with apple trees laden with ripe fruit.

Using his newly acquired conscious mind, Adam thinks, 'I should fly down and munch on some apples'. It seems reasonable but he can't know if it is a good decision or not until he's tried it. He's breezed along, all his stork life, on instinct. Being the only stork with a conscious mind, he can't consult the others and, having just acquired a conscious mind himself, he has as yet no knowledge of what are correct and incorrect understandings.

For his new thinking mind to make sense of the world, he has to learn by trial and error. Having to start somewhere, he decides to carry out his first grand experiment in self-management by flying down to the island and eating some of those delicious apples.

But it's not that simple. As soon as he deviates from his migratory course and heads down towards the island, his instinctive self tries to pull him back on course. In effect, it criticises him for going off course and doesn't want him to search for understanding.

Adam is in a dilemma. If he obeys his instinctive self and flies back on course, he will be perfectly orientated but he'll miss out on the apples. And he'll never learn if it was the right thing to do or not.

All the messages he's receiving from within tell him that to obey his instincts is good. But there's a new message of disobedience, a defiance of instinct. Going to the island will bring him apples and understanding, but it will also make him feel bad.

Uncomfortable with the criticism his conscious mind or intellect receives from his instinctive self, Adam's first response is to ignore the apples and fly back on course. This makes his instinctive self happy and wins approval from the other storks. Not having conscious minds, they are innocent, unaware or ignorant of a conscious mind's need to search for knowledge. They are obeying their instinctive selves by following the flight path past the island.

But, flying on, Adam realises he can't deny his intellect. Sooner or later he has to find the courage to master his conscious mind by carrying out experiments in understanding. (It is only by trial and error or by learning from others' experiences that any of us learn to understand.)

So Adam decides to continue with his experiments in self-management. This time he thinks, 'Why not fly down to the island and have a rest?' Not knowing any reason why he shouldn't, he goes ahead with the experiment. But again his instinctive self criticises him for going off course.

This time he defies the criticism and perseveres with his experiments in self-management. But it means he has to live with the criticism. Immediately he is condemned to a state of upset. A battle has broken out between his instinctive self, perfectly orientated to the flight path, and his emerging conscious mind, which needs to understand why it's the correct path to follow. <u>His instinctive self is perfectly orientated, but he doesn't understand that orientation.</u>

What Adam needs to do is appease his innocent, instinctive self with some vital information. He could begin with the difference between a gene-based learning system and a nerve-based learning system. He should explain that the genetic learning system, which gave him his instinctive orientation, is a powerful learning tool. Over generations it is able to adapt species to new circumstances. But he should also explain that only the nerve-based learning system can learn to understand cause and effect. The problem is that he knows none of this. It is going to take humanity 2 million years to achieve that understanding. In the meantime the battle within him continues at a furious pitch.

To see how instincts are only <u>orientations</u> and not <u>understandings</u> we could consider how instincts are acquired.

The genetic make-up of birds includes a response to the Earth's magnetic field and other factors useful in direction finding. And though they are all similar, their response varies slightly just as they vary slightly in their looks.

Suppose that while the storks were in Africa a volcano erupted in their flight path off the coast of Africa, forming a large island mountain too high to fly over.

When the storks reach the island on their migration, their differing genetic make-up causes some to fly east around it and others to fly west. The latter perish because that route happens to take them too far around the inhospitable island. The others survive and perpetuate the genetic make-up that makes them fly east around the island.

Species can be genetically adapted, or oriented to new circumstances but orientation is not understanding.

Adam Stork needed to be able to explain that a nerve-based learning system, being memory-based, can understand cause and effect. Once you can remember past events, you can compare them with current events and identify common or regularly occurring experiences. This knowledge of, or insight into, what has commonly occurred in the past enables you to predict what is likely to occur in the future and to adjust your behaviour accordingly.

A nerve-based learning system can associate information, it can reason how experiences are related. It can learn to understand and become conscious of the relationship of events through time. A genetic learning system on the other hand can't become conscious.

When the fully conscious mind emerged, it wasn't enough for it to be orientated by instincts. It had to find understanding to operate effectively and fulfil its great potential to manage life. Tragically the instinctive self didn't 'appreciate' that need and 'tried to stop' the mind's necessary search for knowledge, that is, its experiments in self-management. A battle between instinct and intellect developed.

(All underlinings in quotes and text are my emphasis.)

I spoke to you earlier on of this dark child of nature, this other primitive man within each one of us with whom we are at war in our spirit.
Laurens van der Post, *The Dark Eye in Africa*, 1955.

[Origen argued that] **at the beginning of human history, men were under supernatural protection, so there was no division between their divine and human natures: or, to rephrase the passage, there was no contradiction between a man's instinctual life and his reason.**
Bruce Chatwin, *The Songlines*, 1987.

To refute the criticism from his instinctive self, Adam Stork needed to know the difference in the way genes and nerves process information. But when he diverted to the apple tree, he was only taking the first tentative steps in the search for knowledge. He was in a catch 22 situation. In order to explain himself, he needed the understanding he was setting out to find. He had to search for understanding without the ability to explain why. He couldn't defend his actions. He had to live with criticism from his instinctive self. Without defence, he was insecure in its presence. He had to live with a sense of guilt that wasn't justified.

What could he do? If he abandoned the search he'd get short-term relief, but the search still had to be undertaken.

All he could do was retaliate against the criticism, try to prove it wrong or simply ignore it, and he did all of those things. He became angry towards the criticism. In every way he could he tried to demonstrate his worth — to prove that he was good and not bad. And he tried to block out the criticism. He became angry, egocentric and alienated or, in a word, upset.

> **London, 1970: At a public lecture I listened to Arthur Koestler airing his opinion that the human species was mad. He claimed that, as a result of an inadequate co-ordination between two areas of the brain — the 'rational' neocortex and the 'instinctual' hypothalamus — Man had somehow acquired the 'unique, murderous, delusional streak' that propelled him, inevitably, to murder, to torture and to war.**
> Bruce Chatwin, *The Songlines*, 1987.

Arthur Koestler explains his 'inadequate co-ordination' theory more fully in his prologue to *Janus: a summing up,* 1978, in which he also says **Thus the brain explosion gave rise to a mentally unbalanced species in which old brain and new brain, emotion and intellect, faith and reason, were at loggerheads**.

Adam was in an extremely unpleasant position. He had to endure being upset until he found the defence or reason for his mistakes. Suffering upset was the price of his heroic search for understanding.

A sadder and a wiser man . . .
Samuel Taylor Coleridge, *The Rime of The Ancient Mariner*, 1797.

Consequently, we can see that he was good while he appeared to be bad. Upset was inescapable in the transition from an instinct-controlled state to an intellect-controlled state. His uncooperative or divisive aggression and his ego-centric efforts to prove his worth and evade criticism became an unavoidable part of his personality. Such was his predicament, and such has been the human condition.

The paradox of the human condition was that while we appeared bad we always believed that we were good. Our hope and faith was that one day we would be able to explain this contradiction, liberating ourselves from our sense of guilt.

OK, so the only way that most of us humans can fly is by making a booking with an airline. We don't have a great deal in common with storks, but the point is that all animals, including humans, have an instinctive self. Carl Jung called our common, or shared by all, instincts 'the collective unconscious'.

Jung regards the unconscious mind as not only the repository of forgotten or repressed memories, but also of racial memories. This is reasonable enough when we remember the definition of instinct as racial memory.
International University Society's Reading Course and Biographical Studies, Volume 6, c 1940.

The Tao acts through Natural Law . . .
From ancient times to the present,
Its name ever remains,
Through the experience of the Collective Origin.
From the 21st passage of *Tao Te Ching*, attributed to Lao Tzu
(604–531 BC), as translated by R.L. Wing.

While humanity wasn't orientated to migratory flight paths, we still must have had instinctive orientations to guide us before we acquired consciousness. Importantly, then, what was our main instinctive orientation? Perhaps the answer can be found in the photograph on page 50.

Do you get the feeling you are looking at a single entity rather than 16 individual gorillas? (Yes, there are 16.) There's no distrustful personal space between them; they're harmonious, in fact so secure in each other's presence that there are no furtive sideways glances, they are all looking outward. In their oneness they take each other for granted. Ants in their cooperative or integrated nests and bees in their hives do the same. There's no insecurity or divisiveness.

Certainly studies have shown that there is still some uncooperative or divisive behaviour among gorillas, but having lived with unjust criticism of our own divisive behaviour, aren't we likely to stress any divisiveness in other animals (dumb and innocent as they are) to make us feel better? The films *King Kong* (1933) and *Murders in the Rue Morgue* (1932) seem to wildly exaggerate aggression in gorillas. One way Adam Stork coped with unjust criticism was by evading it. It's understandable that we would evade recognition of cooperative integrative behaviour in other animals if it represented unjust criticism of our own divisive nature.

I confess freely to you, I could never look long upon a monkey without very mortifying reflections.
William Congreve in a letter to Jean Baptiste Denis, 1695. Mentioned in Shirley C. Strum's book (about baboons), *Almost Human*, 1987.

The 'aggressive' label belongs to humans rather than gorillas. In the 1979 BBC television series *Life on Earth*, David Attenborough is seen on a hillside in Rwanda embraced by a band of gorillas. In the only instance in 13 episodes that he brings humans into the dialogue, he says: **If there was any possibility of escaping the human condition and living another animal's life it must be as a gorilla . . . It seems very unfair that man should have chosen the gorillas to represent all that is violent, which is not what the gorilla is — but we are.**

Dian Fossey (who wrote *Gorillas in the Mist*, 1983) was impressed and inspired by their cooperative harmony, their sheer unity or love.

[After being approached and touched by a male silverback mountain gorilla in Rwanda's Volcano Park forest] **I felt a kind of unspecified glow, something within that was very much like love, and it came to me then that for the past fourteen years Dian Fossey had literally lived in that glow.**
Tim Cahill, *A Wolverine is Eating My Leg*, 1990.

If gorillas are similar to our ape ancestors, perhaps this photograph suggests a time before upset, when there was no anger, egocentricity or alienation.

The great frontier between the two types of mentality is the line which separates non-primate mammals from apes and monkeys. On one side of that line behaviour is dominated by hereditary memory, and on the other by individual causal memory . . . The phyletic history of the primate soul can clearly be traced in the mental evolution of the human child. The highest primate, man, is born an instinctive animal. All its behaviour for a long period after birth is dominated by the instinctive mentality. . . . As the . . . individual memory slowly emerges, the instinctive soul becomes just as slowly submerged . . . For a time it is almost as though there were a struggle between the two.
Eugène Marais, *The Soul of the Ape*, written in the 1930s, published in 1969.

In the chapter *How We Acquired Our Conscience* it will be explained how humanity acquired a perfect instinctive orientation to cooperative or integrative behaviour. Before introducing that explanation it is necessary to remove our need to evade certain truths. At this point the objective is to examine the possibility (as suggested by the photo of the gorillas) that humans have a cooperative past.

If our original instinctive self or soul, the voice of which is our conscience, was perfectly orientated to the ideals of being cooperative or integrative when our fully conscious intellect emerged, our intellect would have had to know why cooperative, integrative behaviour was important. But tragically, when our intellect began its search for understanding, our conscience unjustly criticised it.

> . . . **our nature** [conscience — is] . . .
> **A sharp accuser, but a helpless friend!**
> Alexander Pope, *An Essay on Man*, Epistle II, 1733.

> [can] **innocence, the moment it begins to act . . . avoid committing murder** [?]
> Albert Camus, *L'Homme Révolté,* 1951 (Published in English as *The Rebel,* 1953).

Our instinct had no sympathy for the search for knowledge. If it had had its way the search would have stopped. We had no choice but to defy our conscience and suffer its unjust, and thus upsetting, criticism.

It's an explanation that parallels the story of Adam and Eve. Genesis (1:27) tells us that we were **created . . . in the image of God**. We were once perfectly orientated to cooperative, selfless, loving ideals. Then Adam and Eve ate the fruit of the tree of knowledge in order to **be like God knowing** (Genesis 3:5). In short, we went in search of understanding.

Having eaten the forbidden fruit Adam and Eve were in deep trouble. We're told they were evil and cast out of Eden. When we went in search of understanding, our upset nature emerged. Life was no longer a Garden of Eden, but a quest for conscious understanding of our purpose.

> **Take the wonderful story of Adam and Eve, the Garden, the apple, and the snake . . . Is it a story of our fall from grace and alienation from our environment? <u>Or is it a story of our evolution into self-consciousness</u> . . . ? Or both? It is also a story of human greed and fear and arrogance and laziness and disobedience in response to the call to be the best we can be.**
> M. Scott Peck, *The Different Drum*, 1987.

> **One semi-plausible theory** [to explain our loss of sensitivity] **is Julian Jaynes's idea of the bicameral mind** [see Julian Jaynes, *The Origin of Consciousness in the Breakdown of the Bicameral Mind*, 1976]. **According to Jaynes, humankind was <u>once possessed of a mystical, intuitive kind of consciousness</u>, the kind we today would call 'possessed'; <u>modern consciousness as we know it simply did not exist</u>. This <u>prelogical mind</u> was ruled by, and dwelled in, the right side of the brain, the side of the brain that is now subordinate. The two sides of the brain switched roles, the left becoming dominant, about three thousand years ago, according to Jaynes; he refers to the biblical passage (Genesis 3:5) in which the serpent promises Eve that '<u>ye shall be as gods, knowing good and evil</u>'. Knowing good and evil killed the old <u>radiantly innocent self</u>; this old self reappears from time to time in the form of oracles, divine visitations, visions, etc. — see Muir, Lindbergh, etc. — but <u>for the most part it is buried deep beneath the problem-solving, prosaic self</u> of the brain's left hemisphere. Jaynes believes that if we could integrate the two, the 'god-run' self of the right hemisphere and the linear self of the left, we would be truly superior beings.**
> Rob Schultheis, *Bone Games*, 1985.

The demise of the imaginative right side of our brain is explained on page 106.

Throughout our history, theologians, writers, poets and artists have described and represented our predicament (as the story of the Garden of Eden does so well), but they couldn't explain it. And only explanation could clarify the question of our guilt. We needed clear biological understanding.

The discipline of science had to be developed. With it came all the details, or mechanisms, of our world. <u>Science made possible a clarifying, biological explanation of why we became upset</u>. It was only in this century that science achieved understanding of the gene and nerve based learning systems, with which we at last have the means to resolve the riddle of human nature. We can explain that there were two different learning systems, each of which needed to learn about integration in its own way. The result was our upset state or condition. Knowing that genes can orientate but only nerves can understand explains our 'mistakes'. We can now see that we weren't bad or guilty after all, which frees us from our sense of guilt and ends the human condition.

…how can there ever be any real beginning without forgiveness?
Laurens van der Post, *Venture to the Interior*, 1952.

At last we can answer the question of questions that William Blake asked in such simple words in *The Tyger,* quoted in the *Introduction*. Yes, he who made the lamb did make thee. We are not evil beings. We are part of God's plan after all. We are relevant/meaningful/worthy. We can love ourselves and each other now.

<u>Compassion</u> leaves an indelible blueprint of the recognition that life so sorely needs between one individual and another; one nation and another; one culture and another. It is also valid for the road which our spirit should be building now for crossing the historical abyss that still separates us from a truly contemporary vision of life, and the increase of life and meaning that awaits us in the future.
Laurens van der Post, *Jung and The Story of Our Time*, 1976.

To find understanding of our instinctive ideals, we had to battle with the ignorance of our instinctive self or soul. Upset was unavoidable. Our predicament was summed up in *The Man of La Mancha*. We had to be prepared **To march into Hell for a Heavenly cause**. We had to lose ourselves to find ourselves. Christ asked, **What good is it for a man to . . . forfeit his soul?** (*The Bible*, Mark 8:36). The answer is we had to forfeit our soul to find knowledge.

Unable to explain why we had to make mistakes, we became aggressive towards the unjust criticism. We tried to prove it wrong and to block it out of our minds. We became angry, egocentric and alienated. Upset or 'sin' was born. As well we sought material comfort and self-aggrandisement to give ourselves the glory we knew we deserved, glory denied us by our soul's ignorance of our true or greater goodness.

Having become upset because we couldn't defend our actions, it follows that finding the defence eliminates upset. Understanding our true goodness is the key to real peace on Earth. Adam and Eve were heroes, not villains. Humanity was unjustly banished from the Garden of Eden. We lost innocence because we appeared bad and couldn't explain that we weren't.

While our instinctive self or soul became perfectly orientated to the cooperative ideals it didn't understand those ideals. When our conscious mind went in search of understanding of those ideals our instinctive self criticised the search which made us angry, egocentric and alienated. This divisive behaviour attracted further criticism from our soul because it expected us to behave cooperatively or integratively. But we can now see that our divisive, 'corrupt', 'evil', 'unsound', upset state was unavoidable. Our soul's unjust criticism caused our so-called 'fall from grace'.

> **Oh wearisome Condition of Humanity!**
> **Borne under one Law, to another bound:**
> **Vainely begot, and yet forbidden vanity,**
> **Created sicke, commanded to be sound:**

What meaneth Nature by these diverse Lawes?
Passion and Reason, selfe-division cause:
Fulke Greville, from his play *Mustapha*, c 1594–96.

Explanation of our fundamental goodness brings with it the possibility of our return to the ideal state. Our upset can subside now that we know we are good and not bad. Our soul's criticism no longer upsets us. We are secure now. We can return to the non-upset ideal state we've longed for, be it called Heaven, Paradise, Eden, Nirvana or Utopia. The difference is we'll arrive in a knowing state. We will be **like God knowing**.

Originally, we were instinctive participants in the ideals. Now we will participate in them consciously and be like gods, knowing, non-upset managers of the world.

We shall not cease from exploration
And the end of all our exploring
Will be to arrive where we started
And know the place for the first time.
T.S. Eliot, *Four Quartets*, from Part 5 of *Little Gidding*, 1942.

Humanity's journey has been astonishing. The greatest story ever told is our own.

We no longer need to block out our instinct's criticism of our intellect. For example we no longer need to avoid criticism by emphasising divisiveness in other animals. We no longer need to evade the significance of integration. Now we can acknowledge that evasion of integrative or cooperative ideals has happened on an overwhelming scale. The happy irony is that it has all been for the ultimate strengthening of those ideals.

Humanity has had to cope this far by being evasive but now this evasion can end and many hidden truths can be revealed.

I will utter things hidden since the creation of the [upset] **world.**
The Bible, Matt. 13:35.

Science and Religion

(The word 'science' comes from the Latin *scientia* which means knowledge. 'Religion' comes from the Latin *re-ligare* which means 'to bind' or integrate.)

Is there no meaning to life?

'What is the meaning of life?' This question has no answer except in the history of how it came to be asked. There is no answer because *words* have meaning, not life or persons or the universe itself.
Julian Jaynes (Professor of Psychology at Princeton University, USA), *Life*, December 1988.

Or is there meaning?

We seem to be on the verge of discovering not only wholly new laws of nature, but ways of thinking about nature that depart radically from traditional science.

Way back in the primeval phase of the universe, gravity triggered a cascade of self-organizing processes — organization begets organization — that led, step by step, to the conscious individuals who now contemplate the history of the cosmos and wonder what it all means.

There exists alongside the entropy arrow another arrow of time, equally fundamental and no less subtle in nature . . . I refer to the fact that the universe is *progressing* — through the steady growth of structure, organization and complexity — to ever more developed and elaborate states of matter and energy. This unidirectional advance we might call the optimistic arrow, as opposed to the pessimistic arrow of the second law.

There has been a tendency for scientists to simply deny the existence of the optimistic arrow. One wonders why.
Paul Davies (Professor of Mathematical Physics at Adelaide University, Australia), *The Cosmic Blueprint*, 1987, from Chapters 10, 9 and 2 respectively.

Science has concerned itself with finding understanding of the mechanisms of existence. It has been reductionist, avoiding the overview; mechanistic, not holistic. The *Concise Oxford Dictionary* defines holism as **the tendency in nature to form wholes . . .**

Holism is a confirmation of the development of order, or integration of matter. But it's a truth we've had to evade because acceptance of integrative meaning unjustly condemned our unavoidable divisiveness. We have been competitive, aggressive and selfish, not cooperative, loving and selfless. Adam Stork coped with unjust criticism by blocking it out or evading it. Now that our divisiveness is defended we can allow ourselves to recognise the truth that matter forms wholes or integrates.

An unmistakable characteristic of matter is its tendency to form more stable and ever larger or more complex wholes. Mathematicians and physicists are still debating the origin of the universe. The 'big bang' theory is currently the most popular explanation, but whatever happened in the beginning, time, space and the fundamental particles of matter, the building blocks of our universe, came into being. The first primitive components, mainly helium and hydrogen atoms, aggregated to form stars, star clusters and eventually galaxies. Inside the stars other complex nuclei and then

41

atoms formed and were spewed out in cataclysmic supernovae to later become planetary systems. Matter continues to form in the thermonuclear reactions within the stars but importantly, the pattern and direction of this genesis is always the same; from the simple to the more complex.

The next level of complexity or order occurred when atoms assembled to form molecules and crystal lattices. Then compounds of various elements came together or integrated to form more complex materials. The greatest breakthrough in the development of order finally came when a complex compound called DNA (deoxyribonucleic acid), which happened to be able to copy or replicate itself, was formed. This was highly significant because suddenly there was a complex substance that could not only perpetuate itself, but by virtue of subtle variations in its molecules, could both adapt to changing circumstances and develop, refine or find ways to achieve even greater order. DNA had the ability to develop larger (in space) and more stable (in time) systems of matter. The DNA unit of inheritance is called a gene and the study of the process of change that genes undergo is called genetics. Genes are tools for developing order.

Genetic development (or genetic refinement) is an integrative process that develops or refines greater order. It is not, as mechanistic science has had to evasively maintain, a divisive process. Genetic refinement is not about competition for survival, it's about developing order. The DNA replicate that persisted was the one that was more stable and able to develop greater order. The search was for order. The existence of division and divisive behaviour indicates incomplete development, not that the purpose of existence is to be divisive. Genetic refinement is an information processing or learning system and what it is learning or creating or refining or developing is greater order of matter.

Subject to the influence of the laws of physics, matter is developed or refined into larger, in space, and more stable (durable or lasting), in time, arrangements or systems.

Matter integrates. It becomes ordered. That is what happens to the matter, time and space ingredients of our world. The goal or purpose or meaning of existence (for a conscious being) is to develop order. Genes, with their ability to replicate, are a marvellous tool for developing order, for creating larger and more stable wholes, for learning how to integrate matter.

We can now explain what we meant by 'life'. DNA's ability to replicate itself meant that it could defy breakdown and go on to find or develop even greater stability. This property of replication or duplication (called 'reproduction' in single-celled organisms and 'growth' in multicellular organisms), which had the effect of turning a brief lifetime into a relatively indefinite one, was the advent of what we call 'life'. This is now an unnecessary demarcation in the story of development. Life or life-time existed before this in all systems of matter, even those below the development level of DNA, but those lower systems were either relatively simple in the variety of matter involved — simple molecules — or relatively unrefined in their ability to develop the order of matter. For instance, they couldn't replicate. With reproduction, the earliest form of which was asexual, came generations. Reproduction later became sexual because the mixing of genes in mating contributed extra variety for genetic refinement, which speeded up the process considerably.

Through this genetic refinement process or natural selection of more stable and larger systems of matter, compounds eventually integrated to form single-celled organisms which in turn integrated to form multicellular organisms, then societies of multicellular organisms. Each stage was a part of the grand journey to develop maximum stability and complexity, or order.

Integration, or perfect order of matter where all the parts are brought together to work for the maintenance of total order, would be the logical conclusion for this remarkable process. Another way of expressing this idea is that the aim

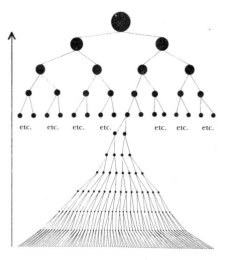

	Integration or harmony of all things (peace on Earth)
	Integration of species (when 'the wolf will lie down with the lamb')
	Societies of single species
	Multicellular organisms
	Single-celled organisms
	Virus-like organisms
	Compounds
	Molecules
	Atoms or the 109 known elements
	Complex nuclei
	Simple nuclei
	Fundamental particles

Development of Order or Integration of Matter

A similar chart appears in Arthur Koestler's book
Janus: a summing up, 1978.

or meaning of life is to achieve universal harmony or 'peace on Earth'.

The essential ingredient of order is that the individuals or parts consider the welfare of the group, or larger whole, above their own welfare. This is unconditional selflessness, the theme of integration and surely what we mean by love.

> **And over all these virtues put on love, which binds them all together in perfect unity.**
> *The Bible,* Colossians 3:14.

Again, the problem with accepting love or selflessness as the meaning of existence and the theme of life is that we would have been confronted by our unjustifiable lack of 'integrativeness'. (Note: since we don't have a word for the process of the development of order or integration of matter

44

I have created the word 'integrativeness' to describe it.)
Integrative meaning condemned us even though deep down
we knew we did not deserve condemnation. This inconsistency forced us to evade any acknowledgement of integrative
meaning, or the development of order of matter.

Science had no choice but to evade recognition of integrative meaning. In its place, it recognised only random
change and termed it evolution. The concept of evolution
was an evasion of the fact of the development of order.
Development can now replace 'evolution'. Now that we can see
we have been part of it, we can openly acknowledge Development.

The discipline of science dictated that it put aside the big
questions of life's meaning, and our inconsistency with it,
and search instead for understanding of the mechanisms
behind the workings of our world. Only with this understanding might we one day discover why we had been divisive and
be able to confront the truth. Without understanding the
mechanisms, we had nothing with which we could explain
ourselves.

Religion has been the custodian of absolute or ideal
truths, safely describing integration or the development of
order as 'God'. It was 'safe' because the term 'God' was
sufficiently abstract not to confront us directly with the
depressing truth of integrative meaning.

Science sensibly turned its back on such dangerous partial
truths as integrative meaning. (While integrative meaning is
an 'absolute truth' it is also a 'partial truth' in the sense that
the full truth, when found, would not criticise us. On its own,
integrative meaning criticises us. As part of the full truth, it
doesn't.) Evading any dangerous partial truths it encoun-
tered along the way, science undertook the tedious search
for rational understanding. Only with such understanding
could the reason for our divisiveness be found and science
and theology, mechanism and holism, reductionism and
vitalism, objectivity and subjectivity, evasive thinking and

45

unevasive thinking be reconciled from their polarised positions.

We can now see a strong implication in religion that 'God' is integrative meaning.

In all theistic religions, whether they are polytheistic or monotheistic, God stands for the highest value, the most desirable good.
Erich Fromm, *The Art of Loving*, 1957.

From the scientific side, there is actually a physical law that explains integrative meaning. As part of my physics course at school I was taught the Second Law of Thermodynamics, which says that everything breaks down to its basic component parts. In scientific terms this means that all energy systems (and matter is a form of energy) must break down until they become heat energy. Since leaving school I have learnt that this law does not apply to 'open' systems which can draw energy from sources outside themselves and which can thereby develop order, or grow in complexity. Earth is an open system, drawing its energy from the Sun. This development of order is known as the 'Second Path' of the Second Law of Thermodynamics. The Second Law of Thermodynamics is also referred to as entropy and its second path as negative entropy.

God is negative entropy, the development of order of matter or, in a word, Development. We can see here that monotheism, the belief that there is only one God, was correct.

God is the laws of physics.
Stephen Hawking (holder of Newton's Chair as Lucasian Professor of Mathematics at Cambridge University) in *Master of the Universe*, shown on the Australian Broadcasting Corporation television program *Quantum*, June 6, 1990.

Science concerned itself with understanding the mechanisms behind absolute truth. It evaded the whole view of integrative meaning by looking only at the mechanisms and by denying the development of order of matter, integration or holism.

Science *had to* deny integrative meaning or God. The Creationist Movement arose to counter that denial. But science had no choice. Only by understanding the mechanisms could we explain why we've been divisive.

> **What has happened in our society in the last half century or so is that our young people in the colleges, universities and schools have been taught the theory of evolution as an established fact. They've been taught that evolution is an exclusively naturalistic theory, and that God is not necessary. God, by definition, is excluded from the process. When the student hears this, he thinks we start with hydrogen gas and our only destiny is a pile of dust.** [The pile of dust is a reference to the evasive emphasis of science on entropy, which implies that disintegration is our destiny or meaning.]
> Dr Duane Gish, Associate Director, Institute of Creation Research, San Diego. *Sydney Morning Herald*, January 8, 1986.

Humanity has been 'God fearing', but now we can confront God. The true role of science has been to liberate humanity from ignorance. The true role of religion has been to comfort humanity while the search went on.

Now we can admit the truth of integrative meaning. We are secure and don't have to cope by blocking out.

> **... I can see a direction and a line of progress for life, a line and a direction which are in fact so well marked that I am convinced their reality will be universally admitted by the science of tomorrow.**
> Pierre Teilhard de Chardin, *Le Phenomène Humain*, written 1938, published 1955 (published in English as *The Phenomenon of Man*, 1959).

While some scientists have begun to break ranks and acknowledge integrative meaning, the proper (i.e. safe) procedure was to find the defence for human divisiveness <u>then</u> admit integration. What was needed was composure before exposure. Rebellious scientists who have acknowledged integrative meaning include professors David Bohm (who wrote *Wholeness and The Implicate Order,* 1980), Ilya Prigogine (*Order out of Chaos,* 1984), Paul Davies (*The Cosmic Blueprint,* 1987) and Charles Birch (*On Purpose,* 1990).

The danger for science was that it would become overly mechanistic/reductionist/evasive, becoming meaningless and morally bankrupt.

The world has achieved brilliance . . . without conscience. Ours is a world of nuclear giants and ethical infants.
General Omar N. Bradley, from his Veterans' Day address, delivered at Boston Massachusetts, November 10, 1948.

It was the appearance of excessive denial in science, not the discovery of the reason/defence for our divisive behaviour, that caused some scientists to abandon mechanism and adopt holism.

Now that mechanism and holism are reconciled they will become an extraordinarily effective and powerful partnership.

Katsushika Hokusai, *Under the Wave Off Kanagawa,* c 1830.

'Do you get the feeling you are looking at a single entity rather than 16 individual gorillas?' (page 33)

'Perhaps this photograph suggests a time before anger, egocentricity or alienation.' (page 34)

Africa — our soul's home — the Garden of Eden.

Fig. 1: Chronology

The fossil history of our human ancestors confirms that rather than branching development (page 55), each stage led to the next, with the sudden emergence of the human condition producing the dramatic change from Childman to Adolescentman. Acknowledging a change at this point, anthropologists renamed the genus from *Australopithecus* to *Homo*.

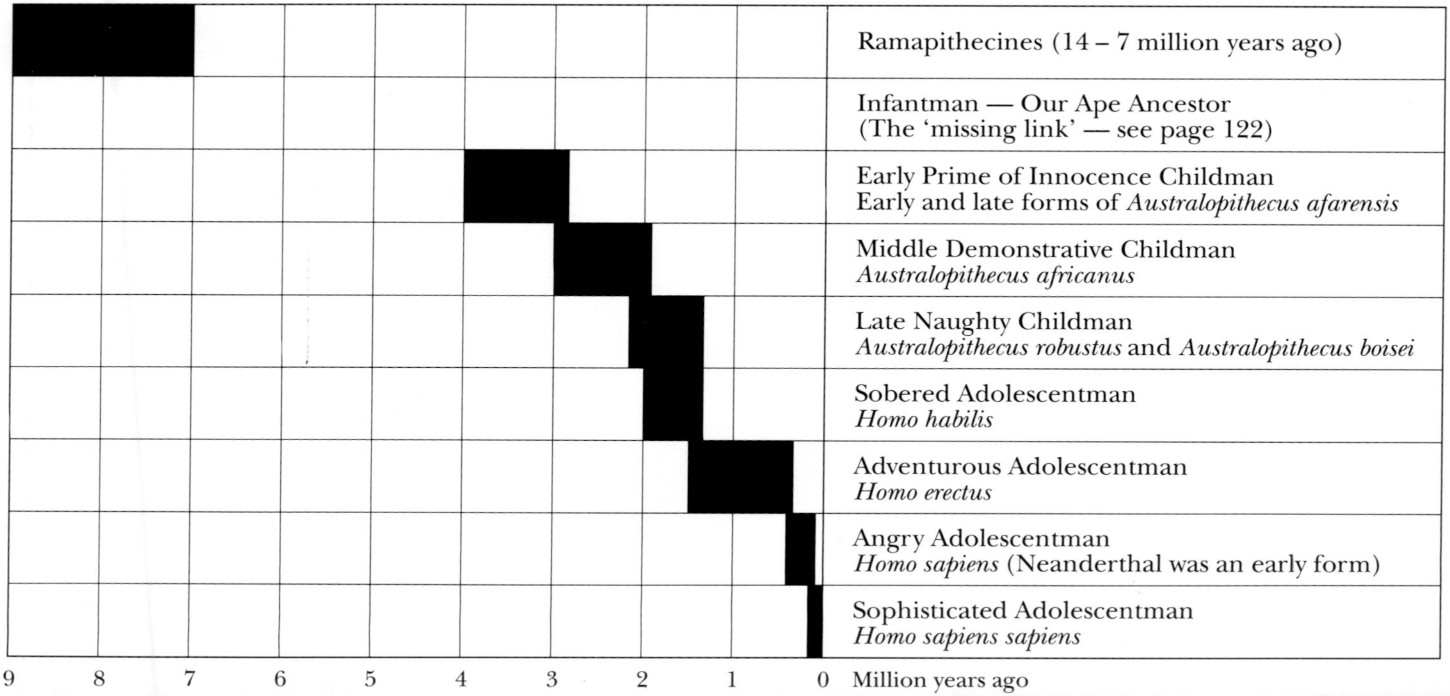

Ramapithecines (14 – 7 million years ago)

Infantman — Our Ape Ancestor
(The 'missing link' — see page 122)

Early Prime of Innocence Childman
Early and late forms of *Australopithecus afarensis*

Middle Demonstrative Childman
Australopithecus africanus

Late Naughty Childman
Australopithecus robustus and *Australopithecus boisei*

Sobered Adolescentman
Homo habilis

Adventurous Adolescentman
Homo erectus

Angry Adolescentman
Homo sapiens (Neanderthal was an early form)

Sophisticated Adolescentman
Homo sapiens sapiens

0   1   2   3   4   5   6   7   8   9   Million years ago

The Story of Homo

A question arising from the story of Adam Stork was 'what was humanity's instinctive orientation?' To suggest an answer the picture of the gorillas was shown. Another question to arise is 'when did the battle between our instinct and our intellect begin?' To answer this consider the fossilised skulls of our ape ancestor Infantman's successors, Childman and Adolescentman. This sequence appeared in *The Search for Early Man* in *National Geographic*, November, 1985. The psychological stages of childhood and adolescence beneath the skulls (my additions) are explained more fully later. For now, suffice to say that, as with the same stages in an individual human, in infancy we became conscious, in childhood we played with the power of free will that consciousness brought and in adolescence we discovered the responsibility of free will and sought the meaning of life so that we might comply with it. When we began the search for meaning the battle with our instinct broke out.

(Traditionally anthropologists have postulated various divergent or branching development in early man, but now we can see that there was no branching — that one stage led to the next. There was only one major development going on, that of the mind. We, unable to look at our psychological development, attributed all significance to everything but

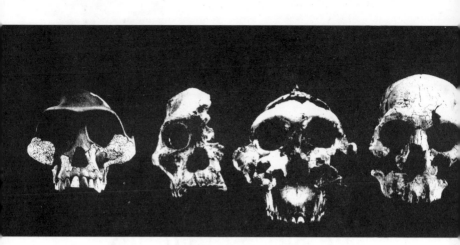

Australopithecus afarensis	Australopithecus africanus	Australopithecus boisei	Human Condition Emerges Here	Homo habilis
Fossil evidence from 4 to 2.8 million years ago	3 to 1.8 m y a	2.2 to 1.3 m y a		2 to 1.3 m y a
Brain Volume 400 cc average	450 cc	530 cc		650 cc
Early Prime of Innocence Childman ➡	Middle Demonstrative Childman ➡	Late Naughty Childman ➡		Sobered Adolescentman

that. The prime mover or main influence in human development was surely not meat eating, tool use, language development or walking upright but what was happening in our minds.)

The chronological sequence of these fossils shows that large-domed skulls appeared 2 million years ago in the genus anthropologists call *Homo*.

Brain volume is probably no longer relevant to intelligence because brain efficiency has no doubt increased, but we may assume that the original emergence of large-domed skulls was to accommodate bigger, more intelligent brains. For example, the developed temporal lobes for remembering and parietal lobes for integrating information from the senses first became clearly evident in the imprint left on the

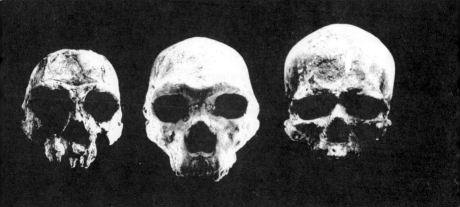

Homo *erectus*	*Homo* *sapiens*	*Homo* *sapiens sapiens*
1.5 to 0.4 m y a	0.5 to 0.05 m y a	0.06 m y a to now
900–1100 cc	1350 cc	1400 cc
➡ Adventurous Adolescentman	➡ Angry Adolescentman	➡ Sophisticated Adolescentman

walls of the brain cases of fossilised *Homo* skulls. As memory and information association form the basis of consciousness, this suggests that Adolescentman *Homo* emerged some 2 million years ago.

———————————

Let's visualise what happened when consciousness, with its ability to understand, emerged. If our instinctive orientation was towards being integrative or loving, what happened when we had to understand love? What happened when our idealistic instinct met the intellect?

To see what happened we can use another analogy, similar to the stork story, but this time featuring humans.

On the 'day' of first contact between instinct and intellect (which in reality happened over many thousands of years), imagine a band of Australos, members of the australopithecines, wandering through the forest and coming upon an apple tree bearing fruit. All the members of this band are still obedient to and dependent on their instincts for management of their lives.

> **Non-rational creatures do not look before or after, but live in the animal eternity of a perpetual present; instinct is their animal grace and constant inspiration; and they are never tempted to live otherwise than in accord with their own . . . immanent law.**
> Aldous Huxley, *The Perennial Philosophy*, 1946.

All that is, except one, whom we'll call Adam Homo. The first fully conscious, intelligent human, he has become sufficiently mentally developed to be able to reason effectively — sufficiently conscious (of how events that occur through time are related) to attempt to think for himself how to behave. For the first time he decides to attempt self-management of his life.

Being an understanding device, his conscious mind requires understanding. But there's none available to help him make decisions. Feeling hungry and having to make a start somewhere on this process of thinking for himself, Adam Homo looks at the fruit and decides, 'why not take all the apples and eat them?'

The rest of the tribe remain obedient to their instinctive training in cooperative or integrative selflessness. They are innocent, unaware or ignorant of the world associated with a search for understanding. Consequently, when Adam Homo takes the apples, they criticise him for not sharing them. (Mothers still witness this grand mistake of pure selfishness made by children when they first attempt to self-manage their lives.) Abused by the innocents and his conscience Adam Homo gets a nasty shock. He quickly puts the fruit down, determined to make no more attempts to self-manage his life.

But he can't deny his intellect. Sooner or later, he has to defy the criticism and shoulder the responsibility of learning to master his intellect.

To stop the criticism, Adam Homo needs to know and explain to the innocents that, while his intellect needs to be guided by conscience, he doesn't deserve its criticism. He needs to explain that he's not bad, that he's using his conscious mind which is a different learning tool from the one that gave them their perfect instinctive orientation to integration.

His mind requires understanding. He has to search for correct understandings by experimenting with different understandings. Any mistakes aren't bad, they're a necessary part of the learning process. But to come up with this explanation, Adam Homo needs 2 million years of investigation. He needs to discover the difference between the gene-based learning system and the nerve-based learning system. He'll need to learn about the DNA molecule and nerves and many other mechanisms of our world before he can free himself from criticism.

We've lived with unfair criticism — on a daily basis and through millennia — and it's upset us. This upset expressed itself as anger, egocentricity, alienation and superficiality.

When the innocents criticised him, Adam Homo's first move was to try to defend his actions. He tried to explain that he didn't deserve criticism, that he wasn't bad. So he said, 'The fruit just happened to fall into my lap'.

This apparently blatant misrepresentation wasn't a lie. It was an inadequate attempt at explanation. Lacking the <u>real</u> excuse or explanation, it was at least <u>an</u> excuse. It was a contrived defence for his mistake. He was evading the false implication that his behaviour was bad. A 'lie' that said he wasn't bad was less a 'lie' than a partial truth that said he was!

The contrived excuses for our divisive behaviour have been Social Darwinism (the need to compete for survival), B.F. Skinner's operant conditioning theory (that man is a slave to reward and punishment), Konrad Lorenz's theory, which excuses our divisive behaviour by saying it is stereotyped and the product of past experiences (i.e. it's instinctive), Robert Ardrey's theory, which said our competitiveness was due to an imperative to defend our territory and Edward Wilson's sociobiology theory, which argues that our selfishness is due to our need to perpetuate our genes.

The latest contrivance is the emphasis given to chaos rather than order in the label 'Chaos' Theory. As Stephen Jay Gould (Professor of Geology and Zoology at Harvard University) said, **Chaos is fundamentally a deterministic theory** (from a lecture in Sydney for the Australian Museum Society, June 15, 1991).

I would point out that each of these contrived excuses contains an element of truth. In fact they represent significant advances in knowledge, but the insight they contain is being presented in such a way as to protect humans from unjust criticism.

The dictionary defines 'ego' as 'conscious thinking self', so ego is another word for the intellect. Trying to find proof of his worthiness, Adam's ego becomes increasingly embattled. It becomes focused or centred on trying to establish his worthiness. He becomes egocentric.

Having to attempt to explain himself also meant Adam Homo had to develop language. The Australos had nothing to explain. They were instinctively coordinated and controlled. Beyond contact calls they had no need for language.

Anthropological evidence supports the belief that language emerged with *Homo*. Study of brain cases in fossil skulls for the imprint of Broca's area (the word-organising centre of the brain) suggests, according to Richard Leakey and Roger Lewin in their book *Origins* (1977), that *Homo* **had a greater need than the australopithecines for a rudimentary language**.

When Adam Homo found he couldn't explain himself satisfactorily, he became frustrated and tried to demonstrate that he wasn't 'bad' or 'inferior' or 'worthless'. He picked up a stick and hurled it away, challenging the innocents to throw one as far. When this sad and desperate effort to demonstrate his worth failed to impress, he retaliated against the unfair criticism. He attacked innocence. He struck one of the innocents in a frustrated attempt to stop the unwarranted criticism. Violence, and in the extreme, war, began.

Our ape ancestor, Infantman, would have occasionally hunted animals to eat, as apes are known to do. Such hunting by apes was graphically illustrated in Sir David Attenborough's 1990 documentary *The Trials of Life* with scenes of common chimpanzees hunting down and eating colobus monkeys.

While Infantman would have occasionally opportunistically hunted animals for food, Childman *Australopithecus* would have been exclusively vegetarian. The trend towards integrative friendliness, even towards other animal species, is clearly evident in the step from common chimpanzees to pygmy chimpanzees (see quotes on chimpanzee behaviour on pages 111 to 120). Unlike common chimpanzees pygmy chimpanzees have developed an instinctive appreciation of love. They have developed conscience. Other evidence, such as reduced canines and the wear pattern on the teeth of *Australopithecus*, indicates that they *were* vegetarian.

While Childman *Australopithecus* would have been exclusively vegetarian, hunting appears again in the life of Adolescentman *Homo*. However, unlike Infantman, who hunted because of the opportunity it offered of acquiring food, Adolescentman 'hunted' (actually attacked) animals because he was upset by their innocence. This will now be explained.

Contrary to accepted opinion, hunting in the hunter-gatherer lifestyle of Adolescentman *Homo* was not for food. In fact research shows that 80 per cent of the food of existing hunter-gatherers, such as the Bushmen of the Kalahari, is supplied by women's gathering. If providing food were not the reason, why were men so preoccupied, as they were, with hunting?

Hunting was upset Adolescentman's earliest ego outlet. Men, who have been more egotistical than women (see page 138) attacked animals because animals' innocence (albeit unwittingly) unfairly criticised them. By attacking, killing and dominating animals men were demonstrating their power in a perverse assertion of their worth. If they couldn't rebut the accusation that they were bad at least they could find some relief from the sense of guilt engendered by demonstrating their superiority over their accusers. The exhibition of power was a substitute for explanation. This 'sport' — attacking animals, which were once our closest friends — was the first great expression of our upset. Anthropological evidence suggests that large-scale big game hunting began during the time of *Homo habilis* and became well established early in the reign of *Homo erectus*. With it came meat eating, which would have revolted our instinctive self or soul since it involved eating our soul's friends. But our spirit (or will to champion our intellect or ego) wasn't to be put off and in time, as we developed our increasingly upset and driven (to find ego relief) lifestyle we became somewhat physically dependent on the high energy value of meat.

The 'hunting for food' explanation is the evasive contrived excuse insecure Adolescentman used for his attack on innocent animals.

As with the !Kung [Bushmen] and other contemporary hunter-gatherers, there was almost certainly more excitement about the men's contribution than the women's, even though the plant foods essentially kept everyone alive. There is a mystique about hunting: men pit their wits and skill against another animal, producing the silent tension of stalking, the burst of energy and adrenalin of the chase, and the elation of success at the kill. The challenge of a hunt is overt and visually impressive, the more so the bigger and fiercer the prey.
Richard Leakey and Roger Lewin, *Origins*, 1977.

Finally, Adam Homo tried to escape the unfair criticism. He 'put his fingers in his ears' to block it out and 'ran away to hide from it'.

'Early vegetarians returning from the kill'

(There is not the same ego satisfaction in
overwhelming a carrot. — J.G.)

Anthropological evidence shows that our first 'migrations' — self-distracting, escapist wanderings — from our ancestral Africa occurred about 1.25 million years ago. For example, the only fossils of pre-*Homo erectus* humanity have been found in the great Rift Valley of Africa. Revealingly, Africa, where humanity was nurtured and spent its innocent youth, is often called 'the cradle of mankind'.

So when humanity went in search of meaning or understanding we became:

- egotistical (forever trying to explain, prove or demonstrate our underlying 'goodness' and thereby maintaining our self-esteem),
- competitive (against the implication that we are bad),
- aggressive,
- mentally 'blocked out' or evasive and thus alienated (paradoxically, we became especially evasive of the selfless ideal, the integrative meaning of life, because it unfairly criticised our selfish, divisive and apparently disintegrative behaviour),
- escapist (superficial), and
- very unhappy.

In short, we became upset.

Since Adam Homo's aggression, competitiveness and preoccupation with self or selfishness were divisive behaviours, they attracted further criticism from his cooperative or integration-demanding conscience. This greatly compounded his war with his instinct and the other Australos. He now needed even more courage and determination to find understanding. Only understanding could free him from criticism. Suddenly there's a desperate need for intelligence to find this understanding.

Anthropologists recognise a dramatic increase in the volume of the human brain beginning 2 million years ago.

Upset intensified quickly. In 2 million years of it we've come a long way from our original innocent state and have almost totally forgotten true happiness. Nevertheless we were once utterly integrative or loving. Our instinctive heritage is one of behaving lovingly towards others and being treated lovingly.

So our forebears were gentle, loving and cooperative, not aggressive, bloodthirsty brutes as has evasively been propounded by such anthropological commentators as Raymond Dart, Konrad Lorenz, Desmond Morris and Robert Ardrey.

From the objective, scientific viewpoint Richard Leakey, who studied and lived amongst the evidence of early man — in fact he and his parents discovered a lot of it — has this to say about the aggression thesis:

> **We emphatically reject this conventional wisdom** [that war and violence are in our genes] ... **the clues that do impinge on the basic elements of human nature argue much more persuasively that we are a cooperative rather than an aggressive animal.**

> **With the growth of agriculture and of materially-based societies, warfare has increased steadily in both ferocity and duration ... We should not look to our genes for the seeds of war** ...
> Richard Leakey and Roger Lewin, *Origins*, 1977, both quotes from Chapter 9.

> **Those who believe that man is innately aggressive are providing a convenient excuse for violence and organized warfare.**
> Richard Leakey, *The Making of Mankind*, 1981.

If the anthropological evidence indicates we weren't aggressive, what might the evidence from the study of primates tell us? Later chapters explain that this evidence too indicates we were not aggressive.

From the subjective, introspective viewpoint our inner awareness of an idealistic, innocent, loving, cooperative past is overwhelming. There is our very deeply ingrained cooperative-behaviour-insisting conscience itself, the idealism of our youth and our expectation to be loved. Where did these instincts come from? A brutish, savage, wild, aggressive, divisive past? Impossible! We even have an instinctive memory that crops up in all mythologies of a time when we were innocent of upset. The fact is our soul, with its undeniable expectations of love, exists in us. No one complains when they are loved! As well, we all know innocence accompanies youth. Anger, egocentricity and alienation came with age. Humanity started out innocent.

Every mythology remembers the innocence of the first state: Adam in the Garden, the peaceful Hyperboreans, the Uttarakurus or 'the Men of Perfect Virtue' of the Taoists. Pessimists often interpret the story of the Golden Age as a tendency to turn our backs on the ills of the present, and sigh for the happiness of youth. But nothing in Hesiod's text exceeds the bounds of probability.

The real or half-real tribes which hover on the fringe of ancient geographies — Atavantes, Fenni, Parrossits or the dancing Spermatophagi — have their modern equivalents in the Bushman, the Shoshonean, the Eskimo and the Aboriginal.
Bruce Chatwin, *The Songlines,* 1987.

This shrill, brittle, self-important life of today is by comparison a graveyard where the living are dead and the dead are alive and talking in the still, small, clear voice of a love and trust in life that we have for the moment lost. . . . [there was a time when] . . . All on earth and in the universe were still members and family of the early race seeking comfort and warmth through the long, cold night before the dawning of individual consciousness in a togetherness which still gnaws like an unappeasable homesickness at the base of the human heart.
Laurens van der Post and Jane Taylor, *Testament to the Bushmen,* 1984.

Having battled ignorance for 2 million years humans have now almost spent their innocence. Innocence in human life is now normally extremely brief. We have been immensely heroic in our search for understanding but it has left us extremely upset and exhausted in soul.

They give birth astride of a grave, the light gleams an instant, then it's night once more.
Samuel Beckett from his play *Waiting for Godot,* 1955.

'Dear little swallow', said the Prince. 'You tell me of marvellous things. But more marvellous than anything is the suffering of men and women. There is no mystery so great as misery. Fly over my city, little swallow, and tell me what you see there'.
Oscar Wilde, *The Happy Prince,* 1888.

From Adam Homo's first unknowing mistake developed what the innocents and his instinctive self saw as deliberate mistakes. They saw his upset as proof of his badness. As far as they were concerned, he had become 'evil' or 'sinful'.

It was the instinctive self's ignorance of the conscious mind's need to search for understanding that led to this intolerable situation. Adam Homo was forced to live with a sense of guilt that was completely unjustified and unwarranted.

The human condition is our predicament of having to live with a sense of guilt. Understanding dissolves the guilt: the origin of all human upset was our instinctive self's unjust criticism of our conscious mind's necessary efforts to find understanding.

Summary of the Concept

THE unspeakable suffering, misery, torment, anguish and damage that we have inflicted on ourselves and our fellows on this planet — the animals, plants, soil, sea, rock and air — were unavoidable consequences of our conscious mind's or intellect's necessary search for knowledge.

It helps relieve the guilt we have suffered to know that any species developing a mind from an instinct-controlled state to an intellect-controlled state would have to go through the stages we have been through. Just as a child must grow through infancy, childhood, adolescence, and only then reach adulthood, so our species has had to develop through these stages.

In infancy, consciousness appears and we discover self. In childhood, we play with the power of free will that emerging consciousness brings. In adolescence, the stage humanity has been in for 2 million years, we discover the responsibility of free will and go in search of the understanding we need to successfully self-adjust. It is when we undertake this search that a battle with our instinctive self develops. The nature of the battle is such that, tragically and unavoidably, the instinct makes the intellect feel guilty or insecure. This happens because the instinct is ignorant or unaware of the conscious mind's need to search for knowledge. The result of the battle is that the intellect becomes upset. Unavoidably it becomes angry, egocentric and alienated.

To end the upset adolescent state and enter adulthood the intellect has to find the understanding that explains it is not bad or guilty for searching for knowledge. Adolescence is the time of the search for the intellect's or conscious mind's identity — for understanding of itself; specifically, for the reason it is good and not bad. Finding this understanding, as has now happened, allows the upset to subside. Humanity is about to leave the turbulent adolescent stage and enter the peaceful maturity of adulthood.

To disarm our upset it has been necessary to explain what we have only been able to feel. The expressions of frustrated and, at times, exalted (momentarily relieved) ego associated with the human condition are clear in the photographs overleaf.

We now know the source of our deep-seated egocentricity and anger

I saw [President] **Sukarno** on a pilgrimage to his mother in Eastern Java whom he had not seen for many years. As he knelt at her feet to receive her blessing, I found myself profoundly stirred. More, I . . . was nearly choked by the collective emotion of the crowd of seventy thousand who had followed him and were sobbing with relief because the suppressed and accumulated longing of centuries to see someone of their own kind leading them again now had found a living symbol at last. But again, <u>where did such volcanic forces in the human being come from?</u>

. . . <u>I suspected that we would never know how to set about dealing with these . . . forces until we knew more of their origin, nature and areas of growth.</u>

Laurens van der Post, *Jung and the Story of Our Time*, 1976.

GEOFF BULL

Carlton coach Ron Barassi exalts in his team's victory in the 1970 Victorian Football League grand final, one of the most memorable ever.

Ferdinand Marcos with his wife Imelda defies the world from the balcony of the Presidential Palace after taking the oath of office as president of the Philippines on February 25, 1986. Elsewhere in Manila on the same day, the representative of the People's Revolution, Mrs Corazon Aquino, was also being sworn in as president. The following day Marcos and his family fled the Philippines.

Nazi Rally in pre-war Germany.

PHOTO.: JOHN BURNEY, COURTESY THE WEEKEND AUSTRALIAN, JUNE 23–24, 1984

Sir Robert Muldoon,
New Zealand Prime
Minister and leader of
the right wing National
Party, defies the ideals,
giving anti-apartheid
demonstrators who
confronted him in
Melbourne on
June 22, 1984
'a mark of C-minus:
Could do better'.

Australian boxing champion Jeff Fenech takes a punch from American
Steve McCrory in Fenech's successful 1986 defence of his IBF world
bantam-weight title.

Politics

THE conscious mind had to defy, resist and battle the ignorance of our older and more established instinctive self or soul. To search for knowledge we had to block out the unjustly condemning ideals of our conscience; to experiment in self-management, we had to be <u>free</u> from its oppression.

On the other hand we had to obey our conscience to a degree and avoid too much blocking out of our soul's integrative ideals. If we became too angry, egocentric and alienated, we would become lost from the integrative or cooperative ideals, which would lead to social disintegration. We required guidance from our conscience, even though it unjustly condemned us.

What we needed was a <u>balance</u> between freedom and oppression of intellect (or between repression and liberation of soul). Our 'development' has been our 'progress' towards understanding. Development required that our soul be repressed to some degree, but too much and we would never find understanding. Progress had to be balanced with restraint.

The Statue of Liberty in New York symbolises freedom, but what do we mean by freedom? Freedom from what? To search for understanding we needed freedom from the oppression of the unjustly condemning integrative ideals. The penalty for being free to search for knowledge was that we became angry, egocentric and alienated. As compensation for having to suffer such self-corruption we surrounded ourselves with material comforts. Materialism was the poor substitute for spiritualism — for the ability to explain why we were good and not bad. Money or capital was needed to supply these material rewards. Materialism and capitalism accompanied freedom. As well we sought contrived success through power and glory to satisfy our embattled egos.

There's no more vivid symbol of adventurous, courageous, heroic, defiant, hedonistic and exhausted humanity than the city of New York, where Times Square symbolises life in upset. It's where drug pushers, prostitutes, muggers and beggars work the footpaths while expensive, chauffeur-driven limousines cruise by. America has certainly been **the land of the free and the home of the brave**.

At the other extreme are socialism and communism with their emphasis on being social or communal (i.e. integrative). In these political systems everything but the integrative ideal was denied. As Karl Marx said **The philosophers have only interpreted the world in various ways; the point is** [not to understand the world but] **to change it** (*Theses on Feuerbach*, written in German in 1845). By 'change it' he meant *make* it social or integrated. Socialism and communism had symbols such as the hammer and sickle, the humble tools of work, and big billboard posters showing everyone ideally striving and pulling together. The flaw in such systems was that they were oppressive and lacked incentive. The need to search for understanding was oppressed, as was the need to indulge and pamper ourselves when we became corrupted. Communism and socialism tried to stop thinking and disallow upset. Since neither could be stopped completely, such regimes tended to be unrealistic. Also without incentive or reward for unavoidable upset, boredom and stagnation crept in.

Freedom-stressing capitalism found understanding for the intellect but was corrupting of soul. Communism and socialism fostered the integrative soul but oppressed the intellect. Both positions had good and bad elements.

These were the extreme positions. The idea in democracy was for everyone to vote and a majority opinion to be found. We voted left wing to emphasise socialistic policies if we thought our community was too corrupt, or right wing for free enterprise policies if we thought our community was too oppressed. In this way balance was sought.

We had to repress our conscience while it unfairly criticised our conscious mind. The difficulty in developing understanding was to find a balance between freedom for the intellect and liberation of the soul.

<u>With its criticism of our intellect removed, repression of our soul can end</u>. Our intellect and instinct or conscious and conscience are now reconciled. The left and right wings of politics are now reconciled. Peace comes to all the warring factions.

They will beat their swords into ploughshares and their spears into pruning hooks. Nation will not take up sword against nation, nor will they train for war any more.
The Bible, Isaiah 2:4.

<u>The intellect is the master integrative tool</u>. It's God or Development's greatest invention. It has the potential to 'knowingly' manage the development of order of matter on Earth. With its insecurity at last overcome, it is now free to realise its great potential.

While humans have appeared to be divisive or disintegrative, the full or greater truth is that we are not. At all times we have been committed to integration. We can see now that upset was a necessary part of the search for understanding; each of us had to live with a soul that unjustly criticised and condemned us.

Worse still the sunshine, the rain, the trees and the animals, all things natural, were friends of our original instinctive self or soul. By association, they too criticised us. Nature was not a friend of our apparently bad mind. We have been unjustly unloved and ostracised by nature for 2 million years!

A lull in the battle.

Psychiatry

(The word 'psychiatry' comes from the Greek *psyche*
which means soul and *iatreia* which means
healing — literally 'soul healing'.)

THE most destructive element of the human condition
was the silence, the denial we had to practise. Unable to
explain (defend) our upset state, we couldn't acknowledge
its existence.

To demonstrate this, let's return to Adam Homo who now
has a son, Tom. Since we instinctively expect the world to be
like it was before upset developed, Tom's instinct gives him
certain expectations of a father. He expects an idealistic
world where there is no anger, egocentricity or alienation.
But his father is not idealistic. He is seldom home, always out
trying to achieve a win for his ego. When he does come home,
he is angry. When Tom asks why he is so aggressive and
preoccupied, Adam can't tell him. An unsuccessful attempt
to explain his upset would be seen by an innocent (such as
Tom) as an admission of badness. Since Adam isn't bad his
only options are to say nothing or contrive a false excuse.

Adam's apparent denial of his upset leaves Tom no choice
but to believe his father is bad, so he leaves home. Adam
Homo is rejected by his own offspring. Condemned by his
instinctive self, by nature and by those he loves, his despair
is immense.

How much courage has been demanded of us! We can see now just how marvellous we humans are. We've endured so much injustice for so long it's beyond appreciation — and may be for all time. The real Hell has been here on Earth. Nothing could be worse than making ourselves seem bad, ugly, destructive and wrong to further the cause of constructive good.

For copyright reasons I am unable to reproduce the lyrics of Joe Darian's song *The Impossible Dream* from *The Man of La Mancha* (1965), paraphrased in part as follows:
Dreaming the impossible dream, fighting the unbeatable enemy, coping with unbearable sorrow, running where even the bravest wouldn't dare go and righting unfixable wrongs . . . Being willing to march into hell for a heavenly cause . . . The world will be a better place now that one scorned, scarred man has striven with his last grain of courage to reach a seemingly unreachable star.

<u>Humans are Development's (or God's) great heroes</u>. We are wonderful and sublimely beautiful beings. We have so much to love ourselves for now. This ability to love ourselves and each other is the therapy we have longed for.

What happened in our lives was not the problem so much as our inability to understand why it was happening. If we knew why it was happening we could cope with it honestly. We could avoid becoming upset. Adam's problem wasn't taking the fruit, but his inability to explain why he took it.

Unable to explain their upset, parents could only deny it, give false explanations for it, or simply be silent about it. Kept in the dark about what was going on, children have been left with no alternative but to block out the pain their parents' world of upset caused them.

. . . grown-ups are certainly very, very odd.
Antoine de Saint-Exupéry, *The Little Prince*, 1945.

Since upset first appeared, each new generation has had to learn to repress its true self and adopt prevailing levels of evasion, denial and silence. Now, suddenly this pattern is broken and we can speak the truth.

For copyright reasons I am unable to reproduce the lyrics of Tracey Chapman's song *Why?* (1986), paraphrased in part as follows:

Why do babies starve when there's enough food in the world for everyone? Why are people still alone when there are so many people? Why are weapons that are aimed to kill called peace keepers? . . . Among all these questions and contradictions there are still some who seek the truth. Somebody will have to answer soon. The time is coming when blind people will remove their blinkers and the speechless will say the truth.

Children will no longer have to die inside themselves, adrift on a sea of silence, superficiality and what appear to be lies. They can be told why we are the way we are and be given the ability to reconcile the upset adult world with the ideal world they instinctively expect. Children will now cope with reality. They won't have to resign themselves to a horrible world of evasion of the truth and repression and denial of the magic world of their souls.

The critical psychological point in our lives came when we resigned ourselves to reality. We were born into the world expecting it to be ideal and still like it was before upset appeared, only to discover it wasn't.

Man is born free but is everywhere in chains.
Jean-Jacques Rousseau, *Le Contrat Social*, 1762 (published in English as *The Social Contract*, 1791).

At first we asked why it wasn't ideal. We asked 'Mummy, why do you and Daddy shout at each other?' and 'why are we going to a lavish party when that family down the road is living in poverty?' and 'why is there so much emphasis in the world on sex?' and 'why do men kill one another?' and 'why is everyone so unhappy and preoccupied?' and 'why is everyone so artificial and false?'. Unable to explain their upset world adults could only say something bewildering like 'one day child you will learn life is more complex than it first appears'. In fact we soon learnt that for some reason, no-one wanted to hear these, what are in truth the real questions about human life.

While we stopped asking these questions that adults seemed to find so unsettling we continued to think them. Eventually, normally at about the age of sixteen, the time came when we decided we had to stop worrying about reality and just resign ourselves to it, as everyone else was doing. Usually what finally convinced us of the need to block out ideality was the discovery within ourselves of a depressing lack of idealism. We discovered the human condition without <u>and within</u>.

So we sensibly resigned ourselves to reality. We practised evading the unanswerable questions about the contradictory nature of human behaviour. Eventually the hypocrisy became invisible to us (hence the need, in the *Introduction*, to resurrect realisation of the hypocrisy). By the age of 40 most of us were so skilled at evasion we had lost all recall of the ideal state and saw nothing wrong with the world the way it was.

I don't know many people over, say, 50 who really believe there's a serious threat to the world. Most of the people I know under 30 really believe it.
Australian cartoonist Karen Cooke in the *Sydney Morning Herald*, December 12, 1986.

Incidentally, it can be appreciated here why it is difficult for most adults to read unevasive books. When a post-resignation mind reads information it has been evading, it blocks the information from coming through. Unaware how deeply evasive or alienated we are, we think we find such books difficult to read because of the way they're written. We say the language is too academic or the ideas are not clearly expressed. Christ, who spoke unevasively, understood the problem. He said **Why is my language not clear to you? Because you are unable to hear what I say. . . . The reason you do not hear is that you do not belong to God** [you evade integrative meaning]. (*The Bible,* John 8:43–47).

There have always been conspiracy theories such as the world's rich manipulating events to keep everyone else impoverished and under their control. There is something going on that is not talked about but what is it? What is the great silence on Earth? It is our complete denial of the world of our soul and all the idealistic truths and beauty that reside there. We have had to cope by repressing our soul and, unable to defend/explain what we were doing, we couldn't talk about it. We are alienated but for most of the time we don't admit it, even to ourselves. It is in our humour that the truth comes out. Our falseness is so transparent sometimes it's funny and makes us laugh. (Note here that humour emerged with the human condition.) This book breaks this great silence on Earth; it flouts the rules of evasion. Coming from the unresigned, unevasive, unalienated world, it's looking on our alienated world from outside. Because it's unevasive it leaves most adults in shock and unable to respond to it.

We have had to be evasive because we couldn't defend ourselves but having found our defence, I know I have the right and most importantly, the responsibility to break the silence. While the information in this book is innocent unevasive thought it has reached all the way to the full truth — it doesn't criticise us as innocent thought almost always has done in the past — it is not another naive condemnation of humanity.

It was mentioned above that we don't admit that we are alienated even to ourselves. We forget that we once abandoned idealism and resigned ourselves to realism. Proof of the extent of this self-deception is the great numbers of people who go in search of the truth after resignation, often later in life or after having become born-again to the ideal world. If we realised how determinedly we had been blocking out the truth for virtually all our thinking lives, we would never claim any ability to be able to think truthfully again. To decide not to see and then claim to be able to see is complete hypocrisy. The number of people in recorded history who have been sufficiently innocent to avoid resignation can probably be counted on one hand and yet there are thousands of books claiming access to the truth!

Again, the exciting prospect now is that children will not have to resign themselves to reality. The most effective way to preserve the soul is not to let it go under in the first place. Avoiding resignation, they will retain the ability to think truthfully and thus effectively and, able to stay alive inside themselves, they will retain all their youthful happiness forever. They will be like gods compared to us, but doesn't the world need them! The strength they'll derive from their happiness will produce a zest for living and enthusiasm that will easily solve any remaining problems. Compared with the struggle against the intolerable burden of undeserved criticism by those who preceded them, the task of future generations will be easy.

The truly great heroes are those who lived during humanity's defenceless adolescence, when the whole world effectively disowned them for their unavoidable divisiveness. For 2 million years, humanity has sent generation after generation against the wall of ignorance. We've been incredibly heroic.

Hokusai's famous painting *Under the Wave Off Kanagawa* (pictured on page 49) says it all. Look closely and you'll see humans huddled in that long boat, which is actually moving forward and already has its bow through the great swell! And there are other boats following! Look at the cold fingers trying to pull them down. What courage!

How We Acquired Our Conscience

To reiterate an important point, genetic refinement is an integrative process, not a divisive one as science has had to evasively maintain. Having grown up in an evasive, mechanistic world we are not used to thinking unevasively, but when we do we will see that as an integrative tool, genetic refinement has certain limitations: for example it normally cannot develop unconditional selflessness among large animals. If an animal were to be born with a genetic inclination to give its life to protect the others in its group for example, that trait would eliminate itself when the animal gave its life (since naturally the animal would lose the opportunity to reproduce). Unconditionally selfless traits are self-eliminating, and don't tend to carry on in the species.

Not being able to develop unconditional selflessness or love inhibits the integrative process, because love is the essence of integration. The essential ingredient of integration or order is that the individuals or parts consider the welfare of the group or whole above their own welfare.

Unconditional selflessness is what makes the human body work so well. Every cell in our body has submerged its individuality to the needs and functioning of the larger

83

system, which is our body. Put simply, selfishness is divisive while selflessness is integrative. Selflessness or love is the theme of integration or Development. It is the ideal state.

The problem for the development of order of matter (or 'God' if we like to personify Development), was how to generate unconditional selflessness among large animals, given this genetic barrier against it. The solution was 'love-indoctrination', a remarkable process that produced the first fully integrated species society or group of large animals. The genus in which this occurred was our immediate predecessor, Childman *Australopithecus*. In producing this integration in our forebears, love-indoctrination gave us our conscience and, fortuitously, liberated conscious thought in us.

To explain the process of love-indoctrination: while unconditional selflessness or altruistic behaviour normally cannot be developed genetically in large animals, some of their behaviour appears to be altruistic. For example a female appears to be behaving altruistically when, as sometimes happens, she loses her life protecting her young. In fact the 'sacrifice' is not selfless but selfish. Since the offspring carries her genes, the mother's protective maternalistic genes are ensuring their own perpetuation when she protects the offspring. Genetic traits have to be selfish if they are to carry on in the species. (As stated, this is simply one of genetic refinement's integrative limitations — it does not indicate that the meaning of existence is to be selfish as the evasive theory of sociobiology teaches.)

Importantly, while maternalism is genetically selfish behaviour, in appearance it is selfless behaviour. A mother gives her offspring nourishment, protection and shelter for apparently nothing in return. It was the appearance of selfless behaviour that gave Development the opportunity to generate unconditional selflessness.

A brain is an observer of its world. To the brain of an infant its mother's maternalism appears to be selfless behaviour. This means that while an infant is being nurtured its brain is being taught that the way to treat others is selflessly or lovingly. Further, the longer the infancy the more thorough that training will be. If selflessness is taught sufficiently well in infancy, the adult will behave selflessly.

The 'trick' to love-indoctrination is that while maternalism is genetically selfish it trains the infant brain in selflessness. The longer the infancy and the more thorough the training, the more the species will practise selflessness or love or cooperation as adults, and the more integrative groups of that species will appear. After a while the process of 'genes following the brain' will (as will be explained) reinforce this process and make love an instinctive expectation.

To produce unconditional selflessness and with it integration, all Development needed was a species in which infancy could be prolonged. In most species infancy has to be kept as brief as possible because of the infant's extreme vulnerability to predators. Zebras for example have to be capable of independent flight almost as soon as they are born, which gives little opportunity for them to be trained in selflessness.

Development had to 'find' a species that was capable of looking after the helpless infant. The successful candidates were the primates. Already semi-upright from having lived in trees where they swung from branch to branch, their arms were partially freed from walking and thus available to hold a helpless infant — the critical factor in developing unconditionally selfless behaviour. (Marsupials can support extended infancy but the pouch is like an external womb, allowing little behavioural interaction between mother and infant. It's the 'selfless' <u>treatment</u> that trains the infant in selflessness or love. Also marsupials have to spend most of their time grazing. There is relatively little time for social interaction between mother and infant and thus training in love.)

With the facility in primates to care for the infant, all that remained was to select for a longer infancy and more maternal mothers and integration would develop. Maternalism became much more than mothers protecting their young, it became a case of mothers actively loving them. Significantly, we talk of 'motherly love', not 'motherly protection'.

Of course, ideal nursery conditions were also required — ample food, comfortable conditions and security from external threats — to allow the loving of infants to take place. Infants are extremely helpless and vulnerable, as anyone who has held a baby in their arms will be aware, which means it is difficult enough just to look after them. However, it wasn't sufficient just to look after them, they had to be loved. While extending infancy *was* 'all' that was needed to develop love-indoctrination it was an extremely 'difficult' development even for primates. The nursery conditions had to be absolutely ideal. As well, we have to remember that delaying maturity, as love-indoctrination does, postpones the addition of the new generations that are so vital for the maintenance of species which are limited mostly to single-offspring births. New generations ensure variety. We can see that developing love-indoctrination was not easy which, incidentally, is why many primate species remain stranded in infancy, unable to progress to childhood. Only one part of our ape ancestors' probable range appears to have provided sufficiently luxurious conditions for love-indoctrination to develop — the aptly described 'Cradle of Mankind' — the Rift Valley of Africa.

Once individuals appeared that were trained in selflessness, the genes would follow the training in love, 'reinforcing' it. While at times an expression of selflessness could still mean elimination of the individual involved, selfless behaviour was now appearing in the species in spite of such losses. Similarly, when the conscious mind emerged later and went its own way, the genes followed, reinforcing what was happening. Generations whose genetic make-up in some way or other helped them cope were selected naturally, making our exhaustion somewhat instinctive in us today. We have been 'bred' to survive the pressures of the human condition. The genes would always naturally follow and reinforce any development process, in this they were not selective. The difficulty was in getting development to occur, not in making it instinctive because that was automatic. <u>If it were not for the ability of genetically selfish maternalism to train infants in selflessness, selflessness would not have occurred to be reinforced</u> genetically because selflessness is a self-eliminating trait.

Importantly, the training in selflessness or love was only an orientation, not an understanding. The infant was being 'brain-washed' with love. It was being <u>love-indoctrinated</u>. This perfect instinctive orientation to integration, now much repressed by our upset, is what we refer to as our conscience. Our conscience is that part of our instinctive self or soul that is concerned with how we behave — it's our moral sense.

Science, being evasive of integrative meaning, couldn't acknowledge the integrative 'love-indoctrination' process and actively evaded it, claiming maternalism was nothing more than mothers protecting their helpless infants. We have had to live with the unbearable fact that during the 2 million years of humanity's upset adolescence (preoccupied with the battle against ignorance) we have been unable to love our infants as much as we did during humanity's infancy and childhood. We made it bearable by evading all significance of nurturing in our species' history and in our personal upbringing. With our upset now defended, and our inability to nurture understood, the truth of the significance of nurturing in our lives and in our species' history can be admitted safely at last. We no longer have to blame our upset evasively on genes (or 'nature' as the contrived genetic excuse was called in the 'nature versus nurture' debate) or on 'chemical causes' as has been fashionable.

> Counsellor: He's a very bright, very aware, extremely tense little boy who is only likely to get tenser in adolescence. He needs some special attention.
> Karen: It's because he was first.
> Counsellor: Hm?

Karen: It's because he was our first. I think we were very tense when Kevin was little. I mean, if he got a scratch, we were hysterical. By the third kid, you know, you let him juggle knives.

Counsellor: On the other hand, Kevin may have been like this in the womb. Recent studies indicate that these things are all <u>chemical</u>.

Gil: (points at Karen) She smoked grass.

Karen: Gil! I never smoked when I was pregnant. . . . Will you give me a break?

Gil: But maybe it affected your chromosomes.

Counsellor intervening: You should not look on the fact that Kevin will be going to a special school as any kind of failure on your part.

Gil: Right, I'll blame the dog.

Scene from the 1989 movie *Parenthood.*

Interestingly, as genetic refinement began to support love-indoctrination, our intellect began to support it also. We self-selected integrative traits by consciously seeking out love-indoctrinated mates, members of the group who had had a long infancy and were closer to their memory of infancy (that is, younger). The older we became the more our infancy training in love wore off; we began to recognise that the younger an individual, the more integrative he or she was likely to be. We began to idolise, foster and select youthfulness because of its association with cooperativeness or integrativeness. The effect, over many thousands of generations, was to retard our physical development so that we became infant-like adults. This explains how we came to regard neotenous (infant-like) features — large eyes, dome forehead and snub nose — as beautiful.

The revealing pictures of an infant and adult chimpanzee (from Stephen Jay Gould's book *The Mismeasure of Man*, 1981) show the greater resemblance humans have to the baby, illustrating the influence of neoteny on human development.

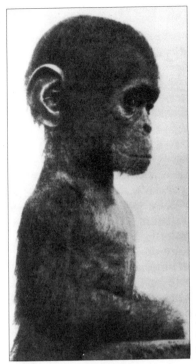

Infant
common
chimpanzee

Adult common chimpanzee

So strong is our attraction to 'cute'/neotenous features, animals that exhibit them, such as seal pups, giant pandas and tree frogs, become favourites.

Would we care if they weren't so cute?
White out the black eye spots and give the ears
points, and the panda loses much of its appeal.

The process of love-indoctrination tells us why and when we lost our body hair.

The interesting fact [is] that man has retained hair chiefly on the scalp, eyebrows, borders of the eyelids, lips and chin, or precisely on those places where hair first appears in all primate foetuses . . .
Adolph H. Schultz, *The Life of Primates*, 1969.

The physical effect of neoteny was that we lost most of our body hair and became infant-looking compared with our adult ape ancestors. We selected for what we now recognise as innocence. (Later, during humanity's upset adolescence, we would become resentful of innocence and instead of

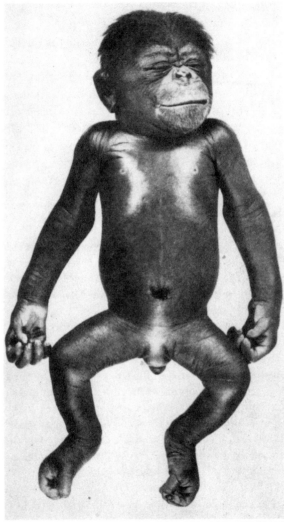

FROM *THE LIFE OF PRIMATES*, BY ADOLPH H. SCHULTZ, 1969

Chimpanzee foetus at seven months showing body hair on the scalp, eyebrows, borders of the eyelids, lips and chin.

cultivating it would seek to destroy it. The attraction of innocence to mate with became perverted. This perversion of the act of procreation is what we now refer to as 'sex', which is explained on page 138.)

Since males were preoccupied competing for mating opportunities before love-indoctrination, females were first to self-select for integrativeness by favouring integrative rather than competitive and aggressive mates. This helped love-indoctrination subdue the males' divisive competitiveness. Without being aware of love-indoctrination, primatologists have noted self-selection of integrativeness by females.

> **Male [baboon] newcomers also were generally the most dominant while long-term residents were the most subordinate, the most easily cowed. Yet in winning the receptive females and special foods, the subordinate, <u>unaggressive veterans</u> got more than their fair share, the newcomers next to nothing.**
>
> **Socially inept and often <u>aggressive, newcomers made a poor job of initiating friendships</u>.**
>
> Shirley Strum, *National Geographic*, November, 1987.

> **The high frequencies of intersexual association, grooming, and <u>food sharing</u> together with the <u>low level of male-female aggression</u> in pygmy chimpanzees may be a factor in male reproductive strategies. Tutin (1980) has demonstrated that a high degree of reproductive success for male common chimpanzees was correlated with <u>male-female affiliative behaviours</u>. These included males spending more time with estrous females, grooming them, and sharing food with them.**
>
> *The Pygmy Chimpanzee*, Edited by Randall L. Susman, 1984, from Chapter 13: *Social Organization of* Pan paniscus *in the Lomako Forest, Zaire*, by Alison Badrian and Noel Badrian.

Note that this self-selection for integrativeness accounts for humanity's rapid development.

We have to explain the speed of human evolution over a matter of one, three, let us say five million years at most. That is terribly fast. Natural selection simply does not act as fast as that on animal species. We, the hominids, must have supplied a form of selection of our own; and the obvious choice is sexual selection. Jacob Bronowski, *The Ascent of Man,* 1973.

The process of love-indoctrination shows that maternalism made us human. Throughout humanity's infancy and childhood, which lasted from 12 to 2 million years ago, nurturing was the most important role in the group. Men had to support it by protecting the group from external threats such as marauding leopards. Women had to devote all their attention to loving their infants if integratively trained and thus cooperative adults were to develop. Until it became instinctive, love-indoctrinated cooperation was extremely hard to develop and maintain.

While humanity was matriarchal (female-role dominated) during its infancy and childhood, it became patriarchal when the threat of ignorance emerged during its adolescence. This patriarchal adolescent stage is explained on page 138.

So love-indoctrination's earliest achievement was the creation of total integrativeness in large animals, namely Childman *Australopithecus.* While we became upset and divisive during our adolescence, we lived utterly integrated lives during our childhood, which lasted from 5 million years ago to 2 million years ago. We were once **in the image of God** (Genesis 1:27) and in Christ's words there was a time when God **loved** [us] **before the creation of the** [upset] **world** (John 17:24) and **the glory . . . before the** [upset] **world began** (John 17:5). **God made mankind upright** [uncorrupted], **but men have gone in search of many schemes** [understandings]. (Eccl. 7:29). Being trained in love as youngsters we practised it as adults and cooperated with each other. We considered the group above ourselves.

> **. . . the basis of all primate social groups is the bond between mother and infant. That bond constitutes the social unit out of which all higher orders of society are constructed.**
> Richard Leakey and Roger Lewin, *Origins*, 1977.

> **But, far more deeply,** [the brain] **depends on the long preparation of human childhood . . . The real vision of the human being is the child wonder, the Virgin and Child, the Holy Family.**
> Jacob Bronowski, *The Ascent of Man*, 1973.

I would mention at this point that, since the understandings being presented represent a reconciliation of science and theology or mechanism and holism, it is to be expected that many of the illustrations for unevasive holistic truths will come from religious teaching. When the metaphysical language of religious teaching is interpreted biologically as it now can be, the soundness or truthfulness of the teaching becomes clear. Truth is where you find it and there is nowhere better to go for the unevasive truth than the great religions. Their teachings are the words of some of the least alienated men in recorded history. (The word 'holy' used to describe true prophets literally means whole or entire; it has the same origins as the Saxon word 'whole', so it confirms the prophets' wholeness or soundness or lack of alienation.) What is more, the teachings of these exceptionally innocent men were recorded before we became highly sophisticated in the art of evading the truth. The fact is, all the great religions are 'gold mines' of unevasive truth. The support people have given them in their millions upon millions over many centuries is evidence of the soundness of their teaching. Christ's words **By their fruit you will recognise them** (*The Bible*, Matt. 7:16) in fact define the scientific method of verification, which depends on reproducible observations and measurements or 'what works'. The predominance of quotes from Christianity reflects in part my greater familiarity with that religion.

94

Pygmy chimpanzees aloft in their forest home along Zaire's Lomako River.

The Rare Pygmy Chimpanzee

Only identified as a species in 1933, pygmy chimpanzees are much gentler and more intelligent than common chimpanzees. Researchers tell of pygmy chimps exchanging prolonged eye contact with them and of the feeling that they try to communicate about things in the past. Their social groups are more stable and they don't tend to cluster in single-sex groups as favoured by their common chimp cousins. In pygmy chimp society the females form alliances and dominate social groups — both male roles in common chimp society. Pygmy chimpanzees have more slender upper bodies than common chimpanzees, are more arboreal, have a greater tendency to walk upright and are known to share their food. While common chimps restrict their plant-food intake mainly to fruit, pygmy chimps eat leaves and plant pith as well as fruit, a diet more like that of gorillas. While they have been known to capture and eat small game they are not known to systematically hunt down and eat large animals such as monkeys, as common chimps are known to do.

With her infant beside her, Kame shares provisioned sugar cane with
Senta, a non-related juvenile, at the Wamba pygmy chimpanzee
research station, Zaire, 1987.

Miso with her infant holds provisioned sugar cane at the Wamba pygmy
chimpanzee research station, Zaire, 1987.

A thoughtful expression occupies the face of the adult male Bosondjo,
as he surveys his fellow pygmy chimps from his perch in an old tyre at
the Yerkes Regional Primate Research Center, Atlanta, GA.

Michelangelo's *The Creation of Adam*
from the Sistine Chapel, 1508-12.

Michelangelo's masterpiece also came to
symbolise our hope of reconciliation with God

How We Acquired Consciousness

A N accidental by-product of love-indoctrination was the liberation of conscious thought. Most species have a block in their minds stopping conscious thought. Before explaining how this block developed and how love-indoctrination was able to overcome it, I should describe the conscious thinking process and explain why we have avoided describing it clearly in the past.

As already noted, the nerve-based learning system, able to remember past events, can compare them with current events and identify regularly occurring experiences. This knowledge of, or insight into, what has commonly occurred in the past enables the mind to predict what is likely to occur in the future and to adjust behaviour accordingly. The nerve-based learning system can associate information, reason how experiences are related, learn to understand and become conscious of the relationship of events that occur through time.

In the brain, nerve information recordings of experiences (memories) are examined for their relationship with each other. To understand how the brain makes the comparisons,

William Blake's frontispiece to *Songs of Experience* (1794)
is an extraordinarily prophetic image of nurturing or
love-indoctrination liberating consciousness.

we can think of the brain as a vast network of nerve pathways onto which incoming experiences are recorded or inscribed, each on a particular path within the network. Where different experiences share the same information, their pathways overlap. For example, long before we understood what the force of gravity was, we had learnt that if we let go of an object, it would usually drop to the ground. The value of recording information as a pathway in a network is that it allows related aspects of experience to be physically related. In fact the area in our brain where information is related is called the 'association cortex'. Where parts of an experience are the same they share the same pathway and where they differ their pathways differ or diverge. All the nerve cells in the brain are interconnected, so with sufficient input of experiences onto a nerve network of sufficient size, similarities or consistencies in experience show up as well-used pathways that have become highways. (In the vast convolutions of our cortex there are about 8 billion nerve cells with 10 times that number of interconnecting dendrites which, if laid end to end, would stretch at least from Earth to the Moon and back.)

An 'idea' describes the moment information is associated in the brain. Incoming information could reinforce a highway, slightly modify it or add an association (an idea) between two highways, dramatically simplifying that particular network of developing consistencies to create a new and simpler interpretation of that information. For example, the most important relationship between different types of fruit is their edibility. Elsewhere the brain has recognised that the main relationship connecting experiences with living things is that they appear to try to stay alive. Suddenly it 'sees' or deduces ('tumbles' to the idea or association or abstraction, as we say) a possible connection between eating and staying alive which, with further experience and thought, becomes reinforced or 'seems' correct. 'Eating' is now channelled onto the 'staying alive' highway. Subsequent thought would try to deduce the significance of 'staying alive' and, beyond that, compare the importance of selfishness and selflessness. Ultimately the brain would arrive at the truth of integrative meaning.

Forgetting also plays a part in understanding (the relationship between experiences). Because duration of nerve memory is related to use, our strongest memories will be of those highways, those experiences of greatest relationship. Our experiences not only become related or associated in the brain, they also become concentrated because the brain gradually forgets or discards inconsistencies or irregularities between experiences. Forgetting serves to cleanse the network of less consistently occurring information, preventing it becoming cluttered with meaningless (non-insightful) information.

Our language development took the same path as the development of understanding. Commonly occurring arrangements of matter and commonly occurring events were identified (became clear or stood out). Eventually all the main objects and events became identified. In the case of language they were named. For example those regularly occurring arrangements of matter with wings we named 'birds' and what they often did we termed 'flying'.

Once insights into the nature of change are put into effect, the self-modified behaviour starts to provide feedback, refining the insights further. Predictions are compared with outcomes leading all the way to the deduction of the meaning to all experience, which is to develop integration of matter.

Consciousness is the ability to understand the relationship of events sufficiently well to be able to effectively/ successfully manage and manipulate those events. For example chimpanzees demonstrate consciousness when they stack boxes and climb them to reach bananas tied to the roof of their cage. Consciousness is when the mind becomes effective, able to understand how experiences are related. It's the point at which the confusion of incoming information clears, starts to fit together or make sense and the mind

LILO H

becomes master of change. (Note, it is one thing to be able to stack boxes to reach bananas — to manage immediate events — but quite another to manage events over the long term, to be secure managers of the world. Infancy is when we discover conscious free will, the power to manage events. Childhood is when we revel in free will and adolescence is when we encounter the sobering responsibility of free will.)

Consciousness has been a difficult subject for humans to investigate, not because of practical difficulties in understanding how our brain works as we're told but because we did not want to know how it worked. We have had to evade admitting too clearly how the brain worked because admitting information could be associated and simplified — admitting to insight — was only a short step away from realising the ultimate insight, integrative meaning, immediately confronting ourselves with our inconsistency with that meaning. Better to evade the existence of purpose in the first place by avoiding the possibility that information could be associated. For the same reason we evaded the term 'genetic refinement' preferring instead the vaguer term genetics. We had to evade the possibility of the refinement of information in all its forms. Admitting that information could be simplified or refined was admitting to an ultimate refinement or law, again confronting us with our inconsistency with that law.

In fact we have avoided not only the idea of meaningfulness but deep, meaningful thinking, which would lead to confrontation with integrative meaning, against which we had no defence. By making deeper insights hard to reach we saved ourselves from exposure but in the process we buried the truth.

Illustrative of our evasion of the nature of consciousness we used the words 'conscious', 'intelligent', 'understanding', 'reason' and 'insight' regularly without ever saying what we are conscious of, being intelligent about, understanding, reasoning or having an insight into, which is how events or experiences are related. The conventional obscure, evasive definition of intelligence is 'the ability to think abstractly'. It was a slip of our evasive guard to name the area of the brain that associates and simplifies information as the 'association cortex'. Of course when we weren't 'on our guard' against exposure few of us would deny that information can be associated, simplified and meaning found. In fact, most of us would say we do it every day of our lives. If we didn't, we wouldn't have a word for 'insight'. But that is the amazing thing about our evasion. We can accept an idea up to a point and then without 'batting an eyelid' go on to pretend it doesn't exist once it starts to lead to a dangerous conclusion.

Our evasion is often obviously false and yet because we have to, we believe it. Look at our evasion of integrative meaning. We are surrounded by integrativeness and yet we deny it. Every object we look at is a hierarchy of ordered matter — an example of the development of order of matter. Our body is composed of parts that are composed of smaller parts, etc. etc. In another example, science doesn't even have a definition for 'love' which is one of our most used words/ concepts. It takes time to become used to the extent of humanity's evasions, our blindness.

In summation, 'insight' was the term given to the nerve highways, the correlation our brain made of the consistencies or regularities it found between events through time. Once we could deduce these insights, these laws governing events in time past, we were in a position to predict or anticipate the likely turn of events. We could learn to understand what happened through time. Our intellect can deduce or distil the purpose to existence or the design inherent in change in information; it can learn the predictable regularities or common features in experience.

Now that we can acknowledge the process of associating experiences that consciousness depends on, the significance of the roles of the left and right hemispheres of the brain become clear.

The right side of the brain specialises in general pattern recognition while the left specialises in specific sequence recognition. The right is lateral or creative or imaginative while the left is vertical or logical or sequential. The right stands off to 'spot' any overall emerging relationship while the left goes in and takes the heart of the matter to its conclusion. We need both because logic alone could lead the intellect into a dead-end. For example, we can imagine that at one point the most obvious association the brain attributed to fruit would have been that it was brightly coloured. However, with more experience the greatest relevance would have been attributed to their edibility. Similar processes occurred in genetic thinking. Dinosaurs seemed to be a successful idea at one stage but ultimately proved otherwise and 'nature' had to back off and take another approach towards integration, namely mammals. When one thought process leads to a dead-end the mind has to back off and find another way in. It goes from the general to the particular and back to the general, until our thinking finally breaks through to the correct association/understanding.

The first form of thinking to wither during alienation was the imagination, because wandering around freely in the mind soon brought us into contact with integrative meaning and its implied criticism. On the other hand if we got onto a logical train of thought that didn't immediately attract criticism there was a much better chance it would stay safely non-critical. Children have wonderful imaginations that they lose before they become adults because they haven't learnt to avoid free/open/adventurous/lateral thinking. Edward de Bono, who trains people to use their imagination again and who has popularised it as 'lateral thinking', once said **often the pupil who is not considered bright will be the best thinker.** (*The Australian*, March 3, 1975.) Figs. 2 and 3 (pages 130 – 131) show that under the human condition alienation increased with intelligence. (Note: these charts are only intended to approximate the developments they describe.) The truth is that in order to think imaginatively we had to remove the need for evasion/alienation. We had to end the human condition.

There must be a reason why other animals haven't developed full mental consciousness; what is it?

One of the integrative limitations of genetic refinement is that it can't reinforce selfless behaviour. In fact, it actively resists it.

Where I live, I can observe the behaviour of a group of kangaroos. Whenever a female comes into season, all the males pursue her relentlessly, attempting to mate with her. So exhausted do the female and the pursuing males become, eventually they are all almost falling with fatigue, yet still the chase goes on. It is easy to see how this behaviour developed.

If a male relaxed his efforts he would lose his chances of producing offspring. Self-interest is fostered by natural selection (genetic refinement). Genetic selfishness is an extremely strong force in animals. It is clear that there would be no chance of a variety of kangaroo developing that considered others above itself. Unconditional selflessness disadvantages the individual that practises it and advantages the recipients of the selfless treatment (such is the meaning of selflessness). At this stage in development genetic refinement automatically acts against any inclination towards selfless behaviour.

In terms of the development of thought this means that genetic refinement was, in effect, totally against a brain thinking selflessness is meaningful. Genetic refinement resisted altruistic thinking in animals. In fact, it developed blocks in their minds against the emergence of such thinking.

In 'visual cliff' experiments newly born kittens crawl to the edge of a table but won't venture over the edge. Presumably, they have an instinctive orientation against going over cliffs. Any kitten that crawled over a cliff fell to its death, leaving only those that happened to have an instinctive block against such self-destructive practices. Natural selection or genetic refinement develops blocks in the mind against behaviour that doesn't tend to lead to the reproduction of the genes of the individuals who practise that behaviour.

Just as surely as kittens were eventually selected for their instinctive block against self-destruction, so were animals selected with an instinctive block against selfless thinking. The effect of this block was to stop the developing intellect from thinking truthfully and thus effectively.

As was explained in the chapter *Science and Religion* selflessness or love is the theme of existence, the essence of integration, the meaning of life. (Christ made the point when he said: **Greater love has no-one than this, that one lay down his life for his friends** [*The Bible*, John 15:13]). If you aren't able to appreciate the significance of selflessness/integrativeness you can't begin to make sense of experience. It is this block against truthful, selflessness-recognising-thinking in most animals' minds that prevents them from becoming conscious (of the true relationship/meaning of experience).

To elaborate, any animal able to associate information to the degree necessary to realise the importance of being selfless towards others would have been at a distinct disadvantage (in terms of its chances of reproducing its genes). Those that don't perceive the importance of selflessness are genetically advantaged. Eventually a mental block would have been 'naturally selected' to stop the emergence of mental cleverness (at associating information). At this point in development, genetic refinement favoured individuals that were not able to recognise the significance of selflessness. The effect was to keep animals stupid, unconscious of the true meaning of life.

Having evaded integrative meaning and the importance of selflessness, it's not easy for us to appreciate that conscious thought depends on the ability to acknowledge the significance of selflessness. The fact is, our own mental block or alienation is the perfect illustration of, and parallel for, this block in animals' minds. Unable to think truthfully/straight we have been unable to think effectively. Alienation has rendered us almost stupid, incapable of deep, penetrating, meaningful thought. The human mind has been alienated from the truth twice in its history: once when we were like other animals, instinctively blocked from recognising the truth of selflessness, and again in our adolescence (which we are just leaving) when we became insecure about our divisive nature and evaded the significance of selflessness and the truth of integrative meaning.

Quite by accident, love-indoctrination breached the block against thinking truthfully by superimposing a new, truthful, selflessness-recognising mind over the older, blocked one. Since our ape ancestors could develop love-indoctrination they were also able to develop truly selfless integrative thinking. They were free to think properly, soundly, effectively and truthfully and so acquired consciousness (the essential characteristic of mental infancy). Chimpanzees are in mental infancy and demonstrate rudimentary consciousness, making sufficient sense of experience to recognise that they are at the centre of the changing array of events they experience. They are beginning to relate information or reason effectively. Experiments have shown they have an awareness of the concept of 'I' or self. Also, as mentioned earlier, they are capable of reasoning how events are related sufficiently well to know that by stacking boxes and climbing them they can reach a banana tied to the roof of their cage.

How did love-indoctrination overcome the instinctive blocks? What actually took place? At the outset the brain was small, with only a small amount of cortex (where information is associated). In this relatively small brain were instinctive blocks orientating the mind away from deep meaningful/ truthful/selflessness-recognising perceptions. At this stage these small inhibited brains were then love-indoctrinated, so although there wasn't much unfilled cortex available, what was there was being inscribed with a truthful, effective network of information-associating pathways. The mind was being taught the truth and given the opportunity to think clearly, over and in spite of the existing instinctive 'lies' or blocks. At first, with the brain so small, this truthful 'wiring' would not have been very significant but it could be developed.

So the mind was trained or 'brain-washed' with the ability to think in spite of the blocks against it. It had been stimulated by the truth at last. Of course it must be remembered that the emphasis in this early stage of the development of love-indoctrination was on training in love, not liberation of the ability to think, which was incidental to the need for integration.

The development of thought, which had been liberated accidentally, was only gradual. The association cortex didn't develop strongly until thinking became a necessity in humanity's adolescence (when we had to find understanding to defend ourselves against ignorance). The large association cortex is a characteristic of Adolescentman *Homo*.

Traditionally the long primate infancy is said to have developed so infants could be taught survival skills, but the evidence shows that learning wasn't strongly promoted until adolescence — after the extended infancy. Now that we don't have to be evasive we can admit the truth, that the long infancy was solely for the development of integration.

The 'need to learn survival skills' explanation implies that survival was a problem, but there had to be ideal nursery conditions for love-indoctrination to develop, which means an environment free of survival difficulties. For example, love-indoctrination and consciousness are more developed in the pygmy chimpanzee than in the common chimpanzee because of the extra comfort and security of the pygmy chimpanzee's environment.

. . . we may say that the pygmy chimpanzees historically have existed in a stable environment rich in sources of food. Pygmy chimpanzees appear conservative in their food habits and unlike common chimpanzees have developed a more cohesive social structure and elaborate inventory of sociosexual behavior. In contrast, common chimpanzees have gone further in developing their resource-exploiting techniques and strategy, and have the ability to survive in more varied environments. These differences suggest that the environments occupied by the two species since their separation by the Zaire River has differed for some time. The vegetation to the south of the Zaire River, where *Pan paniscus* [pygmy chimpanzee] is found, has been less influenced by changes in climate and geography than the range of the common chimpanzee to the north. Prior to the Bantu (Mongo) agriculturists' invasion into the central Zaire basin, the pygmy chimpanzees may have led a carefree life in a comparatively stable environment.
The Pygmy Chimpanzee, Edited by Randall L. Susman, 1984, from Chapter 10: *Feeding Ecology of the Pygmy Chimpanzees* (Pan paniscus) *of Wamba,* by Dr Takayoshi Kano and Dr Mbangi Mulavwa.

The above would seem to indicate that having to live in more variable and less food-rich environments, common chimpanzees have the greater need for intelligence, but that can only be liberated by love-indoctrination and in fact the pygmy chimpanzee is the more conscious or intelligent of the two species.

Everything seems to indicate that [Prince] **Chim** [a pygmy chimpanzee] **was extremely intelligent. His surprising alertness and interest in things about him bore fruit in action, for he was constantly imitating the acts of his human companions and testing all objects. He rapidly profited by his experiences. . . . Never have I seen man or beast take greater satisfaction in showing off than did little Chim. The contrast in intellectual qualities between him and his female companion** [a common chimpanzee] **may briefly, if not entirely adequately, be described by the term 'opposites'.**

Prince Chim seems to have been an intellectual genius. His remarkable alertness and quickness to learn were associated with a cheerful and happy disposition which made him the favorite of all . . .

Chim also was even-tempered and good-natured, always ready for a romp; he seldom resented by word or deed unintentional rough handling or mishap. Never was he known to exhibit jealousy. . . [By contrast] **Panzee** [the common chimpanzee] **could not be trusted in critical situations. Her resentment and anger were readily aroused and she was quick to give them expression with hands and teeth.**
Robert M. Yerkes (who has been described as the dean of American primatologists), *Almost Human*, 1925, from pages 248, 255 and 246 respectively.

Pygmy chimpanzees are exceptionally integrated and are on the threshold of childhood.

We saw above that Chim **seldom resented by word or deed unintentional rough handling or mishap.** The following

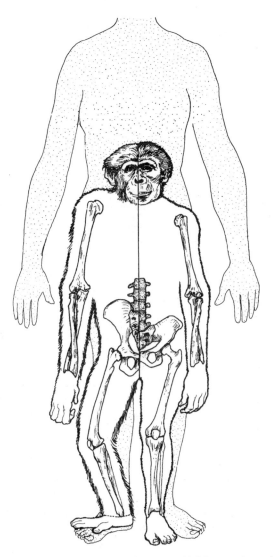

Fossil remains of Early Prime of Innocence Childman, *Australopithecus afarensis* (right) match up remarkably well with the skeleton of a pygmy chimpanzee. The main differences are that *A. afarensis* had a bipedal rather than quadrupedal pelvis, shorter limbs, larger molars and premolars and slightly smaller canine teeth. Pygmy chimpanzees have smaller canine teeth and molars than other apes. A modern human outlined at rear facilitates size comparison.

Drawing by Adrienne L. Zihlman (Professor of Anthropology UC Santa Cruz) from her book, *The Human Evolution Coloring Book*, 1982.

account of Matata, a female pygmy chimpanzee captured in the 'wild', confirms this tolerance, but more importantly, also illustrates the exceptional mothering that love-indoctrination depends on.

> **Kanzi was stolen from his mother (without contest) 30 min after birth by Matata and has been reared by Matata as her son since that time . . .**
>
> **Matata had conceived and borne one infant of her own, Akili . . . Matata cared very adroitly for both infants . . . When . . . 9 months of age, Akili was sent to the San Diego Zoo. Kanzi and Matata were assigned to the language project and remained together . . .**
>
> **Matata and Kanzi are housed in a large six-room indoor-outdoor enclosure** [at the Yerkes Regional Primate Research Center in Georgia, USA]. **Even though Matata was wild-born, both she and Kanzi continually seek out and appear to enjoy and depend upon human companionship. Consequently, human teachers and caretakers are with them throughout the day. Mild distress is evidenced by Matata, even though she is an adult, at the departure of human teachers with whom she has formed close relationships. She seeks to maintain close proximity to these teachers as they work with her, often simply sitting with an arm or leg draped across them. When they stay until evening she pulls them into the large nest that she builds and goes to sleep next to them. Nesting with others, of all sex classes, has been reported in wild *Pan paniscus* and is peculiar to this species (Kuroda, 1980).** [Note that while pygmy chimpanzees *have* been reported nesting together, my inquiries indicate that the usual practice is for each adult to have its own nest.]
>
> **Similar sorts of attachment behavior toward humans are rarely observed in adult apes of other species (Yerkes and Yerkes, 1929). In instances where attachment to humans does occur in other ape species it is invariably preceded by a long period of rearing in which a human becomes a parental surrogate for the ape at a very early age (Patterson and Linden, 1981; Temerlin, 1975). Since Matata came to the project fully adult, with her own offspring to care for, her attachment to the humans around her is clearly not a result of human upbringing.**

Likewise, observations of her social behavior toward other pygmy chimpanzees housed with her at the field station prior to her inclusion in the Language Research Project revealed that she was well integrated into that group. She was the highest ranking female, the male's favorite partner, and a competent mother of two young infants. She was, in general, the focus of group attention and evidenced frequent affiliative behaviors toward all the other individuals in the group.

As a result of these affiliative behaviors, which are extremely similar in form and content to those of humans, it has been possible to introduce new individuals who have no previous experience working with apes to Matata and Kanzi. The ready extension of affiliative behaviors towards new individuals stands in marked contrast to the behaviors displayed by [common chimpanzees who typically] ... display considerable aggressive behavior toward strangers [in] ... both male and female, and in both captive and wild environments ...

Kanzi, like Matata, is extremely affiliative and socially responsive to human interaction and contact. He enjoys being carried around by his teachers and he initiates frequent carrying. Matata not only permits this, but on occasion even encourages Kanzi to go to others by detaching his hands and shoving him in the direction of another individual. This is not a sign of lack of interest in Kanzi, nor a sign of atypical maternal behavior in Matata. While Matata was housed with other pygmies at the field station she also allowed them to carry, play with, and discipline both Akili and Kanzi. She also encouraged these infants to go to other pygmy chimpanzees and retrieved them only when they appeared quite distressed. By one and a half years of age, Akili spent most of his time playing with other individuals in the group, particularly the adult male. He returned to Matata when he was hungry, sleepy, hurt (bumped his head, tripped, etcetera) or when she was moving from one area to another. Kanzi is following a similar pattern of strong attraction to other individuals, though in his case these other individuals are not pygmy chimpanzees, they are humans ...

Interaction with others appears to be initiated primarily (though not entirely) by the strong attraction of young infants toward other individuals. On numerous occasions Kanzi has

leaped directly from Matata's ventral surface onto the body of
an approaching person . . .

As an infant, Kanzi is interested in exploring everything . . .
Although Matata seems to be quite content to have others carry
and entertain Kanzi for long periods, <u>she is always keenly aware
of his location at every moment</u> . . .

<u>Matata is tolerant of a wide variety of interactions between
Kanzi and the teachers</u> . . . [When a knife an experimenter was
using accidently nearly cut Kanzi he] **became furious, scream-
ing and lashing out at the experimenter in attempts to bite. The
experimenter, fearful (not of Kanzi), but that an attack from
Matata would result, looked at Matata with an expression of
dismay and pulled the knife back. Matata, having closely observed the entire set of events, simply pulled Kanzi to her and
tried to quiet him even though Kanzi kept threatening the
teacher. <u>When the teacher reached out toward Matata, Matata
hugged her</u> and again tried to quiet Kanzi. Kanzi tried to bite the
teacher even as his mother was hugging her. He remained angry
for about 20 minutes.**

**It is quite surprising that Matata did not bite the experi-
menter in response to Kanzi's screams. The** *Pan troglodytes*
[common chimpanzees] **females observed by Savage (1975)
would have responded to such an event (intended or not) with
instant aggression.**

The Pygmy Chimpanzee, Edited by Randall L. Susman, 1984, from
Chapter 16: Pan paniscus *and* Pan troglodytes by Dr E.S. Savage-
Rumbaugh.

We can see that Matata's world is above all a secure one
and that she is cultivating that security in Kanzi. As a mother
she appears to be exceptional in her ability to reassure and to
cultivate trust. What she is reassuring Kanzi of, and teaching
him to trust, is love. She has <u>extraordinary trust that love, not
hate, will prevail</u>. The certainty in her behaviour is of the
presence of love. An infant growing up in the common
chimpanzee environment, where **resentment and anger** [are]
readily aroused would not trust that it was not going to be
attacked by others. In such circumstances love or harmony or

integration (evasively referred to in this mechanistic study as **affiliation**) within the group would be tenuous at best. Matata's world is one of extraordinary trust — of belief in and certainty of love — of security of self — of soundness. She is not from a divisive world, but a loving, integrative world and she is cultivating/nurturing that appreciation of love in Kanzi. She is love-indoctrinating him. Note that he, like the common chimpanzee, has not yet acquired a full appreciation of love or conscience.

Matata's security and resulting integrativeness or **affiliative behaviour** is what made her popular and **the focus of group attention.** The others were recognising and favouring integrativeness. This is an example of self-selection.

I might mention here that unlike common chimpanzees, pygmy chimpanzees at times mate face to face and are in general more sexually active than common chimpanzees. Given they are becoming conscious this is not surprising. Consciousness would have brought a greatly enhanced capacity to reflect upon, savour and therefore enjoy pleasant sensations. The point is the increased sexual activity is not due to upset; it is not 'sex' as we know it. The perversion of the act of procreation emerged in Adolescentman as is explained on page 138.

Matata didn't steal Kanzi from his mother out of cruelty, but because of love-indoctrination, which selects for more maternal mothers. In the end it produces a maternal instinct so strong it can become uncontrollable.

The proverbial devotion of wild primates to their young is really exemplary in mothers with the accumulated experience gained from having raised previous infants. That this successful care has benefited also from opportunities to observe other mothers of the group, has been demonstrated by the inefficient maternal behaviour, ranging to completely ignoring or even killing of the newborn, of captive mothers raised in total isolation. The urge to hold an infant, even though not their own offspring,

seems to be irresistible to many female monkeys, according to the common occurrence of 'baby snatching' and adoption among captive macaques and other species. Indeed, there appeared recently a remarkable report of a wild female spider monkey seen to carry an infant howler monkey for several days until the latter died of starvation.
Adolph Schultz, *The Life of Primates*, 1969.

The fact that nurturing requires practice indicates that, intense an instinct as it already is in primates, it is a relatively recent development. The longer a behaviour occurs the more instinctive it becomes. Breathing for example has been practised by animals since they first appeared and is now totally instinctive, automatic or reflex. The potential to love-indoctrinate their infants accompanied the emergence of the primates' 'arms-freed-from walking' arboreal lifestyle.

We can see that pygmy chimpanzees are far more advanced along the love-indoctrination path than common chimpanzees. Gorillas are too, but — I think we will discover — not quite as far advanced as pygmy chimpanzees. Interestingly, while pygmy chimpanzees depended on the safety of trees for the secure, threat-free environment needed to develop love-indoctrination, gorillas 'chose' physical size and great strength for that purpose. To quote Schultz again, the adult male gorilla **is a remarkably peaceful creature, using its incredible strength merely in self-defence**. The following extracts from Dian Fossey's *Gorillas in the Mist* (1983) reveal the strong relationship between nurturing and integrativeness that is love-indoctrination:

Like human mothers, gorilla mothers show a great variation in the treatment of their offspring. The contrasts were particularly marked between Old Goat and Flossie. Flossie was very casual in the handling, grooming, and support of both of her infants, whereas Old Goat was an exemplary parent.
from Chapter 9

The effect of Old Goat's **exemplary parenting** on Tiger (her son) is apparent:

> **Like Digit, Tiger also was taking his place in Group 4's growing cohesiveness. By the age of five, Tiger was surrounded by playmates his own age, a loving mother, and a protective group leader. He was a contented and well-adjusted individual whose zest for living was almost contagious for the other animals of his group. His sense of well-being was often expressed by a characteristic facial 'grimace.'**
> from Chapter 10

The **growing cohesiveness** (developing integration) brought about by **loving mothers and protective leaders** is love-indoctrination.

Dian Fossey's account of the love-indoctrinated Tiger later in his life illustrates how nurtured love is required to produce the integrated group. It describes how the secure, integrative or loving Tiger tried to maintain integration or love in the presence of an aggressive, divisive gorilla after the group's integrative silverback leader, Uncle Bert, had been shot by poachers.

> **The newly orphaned Kweli, deprived of his mother, Macho, and his father, Uncle Bert, and bearing a bullet wound himself, came to rely only on Tiger for grooming the wound, cuddling, and sharing warmth in nightly nests. Wearing concerned facial expressions, Tiger stayed near the three-year-old, responding to his cries with comforting belch vocalizations. As Group 4's new young leader, Tiger regulated the animals' feeding and travel pace whenever Kweli fell behind. Despondency alone seemed to pose the most critical threat to Kweli's survival during August 1978.**
>
> **Beetsme . . . was a significant menace to what remained of Group 4's solidarity. The immigrant, approximately two years older than Tiger and finding himself the oldest male within the group led by a younger animal, quickly developed an unruly**

desire to <u>dominate</u>. Although still sexually immature, Beetsme <u>took advantage</u> of his age and size to begin <u>severely tormenting</u> old Flossie three days after Uncle Bert's death. Beetsme's <u>aggression</u> was particularly <u>threatening</u> to Uncle Bert's last offspring, Frito [son of Flossie]. <u>By killing Frito</u>, Beetsme would be destroying an infant sired by a competitor, and Flossie would again become fertile.

Neither young Tiger nor the aging female was any match against Beetsme. Twenty-two days after Uncle Bert's killing, Beetsme succeeded in <u>killing</u> fifty-four-day-old Frito even with the <u>unfailing efforts</u> of Tiger and the other Group 4 members to <u>defend</u> the mother and infant. . . . Frito's death provided more evidence, however indirect, of the <u>devastation</u> poachers create <u>by killing the leader</u> of a gorilla <u>group</u>.

Two days after Frito's death Flossie was observed soliciting copulations from Beetsme, not for sexual or even reproductive reasons — she had not yet returned to cyclicity and Beetsme still was sexually immature. Undoubtedly her invitations were <u>conciliatory measures</u> aimed at reducing his continuing physical <u>harassment</u>. I found myself strongly disliking Beetsme as I watched <u>his discord destroy</u> what remained of all that Uncle Bert had succeeded in <u>creating and defending</u> over the past ten years . . .

I also became increasingly concerned about Kweli, who had been, only a few months previously, Group 4's most vivacious and frolicsome infant. The three-year-old's lethargy and depression were increasing daily even though Tiger <u>tried to be both mother and father</u> to the orphan.

Three months following his gunshot wound and the loss of both parents, Kweli gave up the will to survive . . .

It was difficult to think of Beetsme as an integral member of Group 4 because of his continual <u>abuse of the others</u> in futile <u>efforts to establish domination</u>, particularly over the <u>indomitable</u> Tiger . . .

Tiger <u>helped maintain cohesiveness</u> by 'mothering' Titus and <u>subduing</u> Beetsme's <u>rowdiness</u>. <u>Because of Tiger's influence</u> and the immaturity of all three males, they remained <u>together</u>.

from Chapter 11

In all ape species there is still a residual amount of uncontained sexual opportunism, such as dominance hierarchy (which orders and thus, to a degree, contains competition for mating opportunities) and infanticide such as mentioned above in the story of Tiger. The newly dominant male will often kill the offspring of his predecessor, bringing the mother back into season earlier than would otherwise occur, allowing the killer to mate and reproduce his genes more frequently. As mentioned, love-indoctrination was not at all easy to develop. It had to overcome some deeply embedded divisive behaviour. Developing love-indoctrination was like trying to swim upstream against a strong current to reach an island. The island was childhood, where love-indoctrination was complete and love had become instinctive. While struggling to reach the island of childhood, Infantman was often swept back downstream. Any breakdown in nurturing would lead to the appearance of divisive adults who had reverted to pre-love-indoctrination competitive, sexually opportunistic practices.

Impasses and difficult stages are natural features of Development as anyone who has tried to develop something knows. Evasive, mechanistic science, unable to acknowledge the purpose in existence of developing order, didn't recognise the occurrence of impasses. If nature is not going anywhere, as the evolution theory maintains, then there can't be any impasses or breaking through them. Unable to recognise impasses, science has not been able to explain stops and starts in nature, the 'punctuated equilibrium' evident in the fossil record. In this instance, the impasse was in integrating primates. Ideal nursery conditions were needed for love-indoctrination to develop, which explains why many primate species are still stranded in infancy. It also explains the fossil void or 'missing link' in the anthropological record of our forebears. We have as yet found no fossil evidence of our ape ancestor, Infantman.

To explain the missing link — 20 million years ago in warm, lush conditions, the forest-dwelling dryopithecines (the prototype apes) emerged. They thrived on fruit, soft leaves and shoots, and were as numerous as monkeys are today. Then about 15 million years ago the world's climate began to cool and the lush forests began to shrink. By then the dryopithecines had developed into the more refined ramapithecines, which flourished between 14 and 8 million years ago and were spread widely across Africa, Asia and Europe in small, medium and large varieties. It is thought that further global cooling and resulting loss of forest habitat around 8 million years ago (when the plains animals of Africa started to proliferate) began to eliminate them. Apparently they were now sufficiently adapted to arms-free walking to begin to develop love-indoctrination, because those that were left almost certainly gave rise to existing ape species and to our ape ancestor, all of whom, to varying degrees, have been able to develop love-indoctrination.

The fossil record from 8 to 4 million years ago is all but a blank for the apes and a complete blank for our ape ancestor. Ape species and numbers are few today, indicating that while they may be receptive to love-indoctrination they are not well adapted to current environmental conditions. Given their rarity today and (presumably) over the past 8 million years, it is not surprising that fossil evidence of them is scarce, but why is there nothing of our ape ancestor? Now that we know how ideal nursery conditions must be to support love-indoctrination, we can appreciate that evidence of our ape ancestor must be even scarcer than that of the other apes. Chimpanzees illustrate the predicament perfectly. Today we have only four varieties of chimpanzee, none of them numerous, with only one, the pygmy chimpanzee, living in an area — and it's only a small area — well suited to developing love-indoctrination. We can imagine that in the far future, fossil evidence of chimpanzees will be even scarcer than it is today, and that of pygmy chimpanzees scarcest of all.

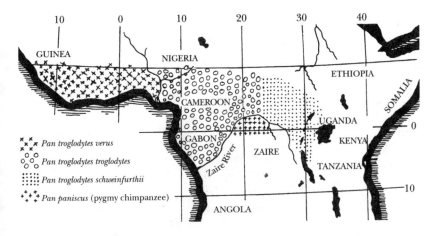

Equatorial Africa,
showing distribution of the four varieties of chimpanzee.
After a drawing in Adolph Schultz's, *The Life of Primates*, 1969.

After reading about nurturing among gorillas and chimpanzees, we can appreciate the difficulty we have had in acknowledging its importance in development. Many mothers will feel awkward as they read this material, because it confronts them with their inability to nurture their children as much as they would like. The truth is, no mother has ever been able to love her child as much as all mothers did before the human condition emerged. It bears repeating that now we can explain/justify our inadequate nurturing, we can safely confront the truth of the importance of nurturing. I should stress that the difference in nurturing ability among apes is due to incomplete love-indoctrination whereas in humans it was caused by our preoccupation with the battle to find understanding.

As mentioned earlier, the pygmy chimpanzee is on the threshold of childhood; in fact I think we could consider Matata a living example of what Childman was like. (Note: it is not of great significance whether pygmy chimpanzees are our genetically closest living relatives — humans and apes

Matata with her adopted son, Kanzi

almost certainly all descended from the ramapithecines —
what is of interest is that pygmy chimpanzees are our psychologically closest living relatives.) Given that pygmy chimpanzees are rare and endangered, that we can now appreciate
how loving and conscious they are and see how much they can
reveal about our own time in 'the Garden of Eden', a con-
certed effort must be made to protect them. I would like the
Foundation For Humanity's Adulthood to establish a fund to
help protect them and to make an unevasive study of them.

It is appropriate to point out again that infancy, childhood and adolescence are the evasive terms we've used to
describe the stages of the emergence of understanding.
Infancy is characterised by self-awareness, childhood by self-
confidence and adolescence by self-understanding or identity search. Adolescence was when the upset human condition developed, the stage humanity is about to leave for the
adult stages of self-implementation, self-fulfilment and self-
maturation.

Once a species has completed infancy, the other stages will
follow naturally. While the rate of development may vary
through childhood and adolescence, unlike what happened
in infancy, development through these stages is inevitable
and (overall) rapid, which is why there are no living examples
of the australopithecines or the stages that preceded *Homo
sapiens sapiens*. In adolescence, the first to descend the exhaustion curve (Fig.3) always replaced the less advanced.
Upset has been replacing innocence for 2 million years,
occasionally through physical conflict but mostly by the more
innocent finding themselves unable to accept and cope with
the new reality, the new level of compromise of the ideals.

Now Abel kept flocks, [stayed close to nature and innocence]
and Cain worked the soil [became settled and began the discipline and drudge of searching for knowledge] . . . **Cain** [be-
came] **very angry, and his face was downcast** [depressed] . . .
[and] **Cain attacked his brother Abel and killed him.**
The Bible, Genesis 4:2,5,8.

125

The dreadful journey through adolescence has made us more and more callous, which means dead in intellect and soul. Thankfully, it is over.

We can see that love-indoctrination has been the most important process in humanity's development. It gave us our conscience and liberated conscious thought in us.

Another consequence of love-indoctrination was that it freed our hands to hold tools and implements and for an almost endless list of other applications. The more love-indoctrination developed and the longer infants were kept in infancy and the more they had to be held, the more we had to stay upright in order to hold them. This freedom of our hands from walking proved extremely useful later when the intellect needed to assert itself, because it could direct the

Rhesus monkey with infant. This picture illustrates the difficulty of carrying an infant and suggests the reason for bipedalism.

126

hands to manipulate objects. A fully conscious mind in a whale or a dog would be frustrated by its inability to implement its understandings.

The following transcript provides some indication of the leisure, fellowship and friendliness that existed among humans during humanity's integrative childhood.

Q: What does your average study day involve?

A: We usually get up before dawn and go out to the nests where the pygmy chimpanzees have slept. We are then there when they get up in the morning. They usually get up very slowly and if it's raining, they like to stay in bed for a long time, which I can understand. They will then often head for a fruit tree that's nearby and they'll have a big feeding bout first thing in the morning before they settle down for a long rest or play period [love-indoctrination period] in mid-morning to midday. They'll then start to travel more, maybe to visit some other fruit trees and this is often when it becomes difficult to follow them. Although if they're moving short distances they go through the trees, when they're going to go a long way they come down to the ground and move very fast and we often end up losing the animals when trying to follow them, but we also know that they know that we're there. The animals that we're most used to following have now got to the stage where they'll come to the ground and start moving off and we move off after them. But we lose them and don't know which way to head, so we'll sit down and wait for them to call and several times now I've been sitting waiting for the animals to call to know which way to go and I'll turn around and there's this face peering at me through the underbrush and it's the pygmy chimps come to see why I've stopped following them.

Dr Frances White (interviewed by Miles Barton), BBC *Science Magazine*, 1990.

The Infant-Like Features of Adult Pygmy Chimpanzees and Gorillas

The more infant-like or neotenous rounder head and eyes, smaller ears, less protruding jaw and less prominent brow ridges of the pygmy chimpanzee and the gorilla indicate that they are more love-indoctrinated than the common chimpanzee.

Infant male common chimpanzee, Gombe.

PHOTO : MICHAEL RAYNER, COURTESY GOOD WEEKEND

Adult female common chimpanzee, Sutu.

TARONGA ZOO

Adult male
pygmy
chimpanzee
in the
San Diego Zoo.

Adult male
mountain
gorilla, Digit.
This is Dian
Fossey's gorilla
friend in
whose memory
the Digit
Fund, Inc., was
established.

Fig. 2: The Development of Mental Cleverness

(Brain volume is used as a guide to mental cleverness)

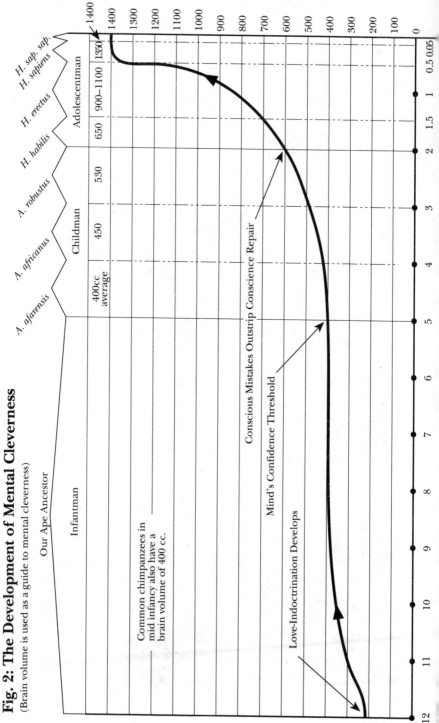

Fig. 3: The Development of Integration

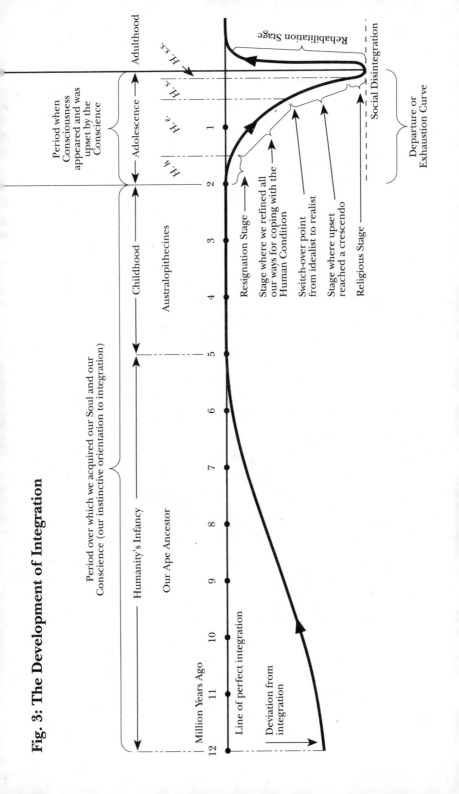

Fig. 4: The Development of Humanity

(Portrayed as a journey via passes through mountain ranges that bar the way to integration.)

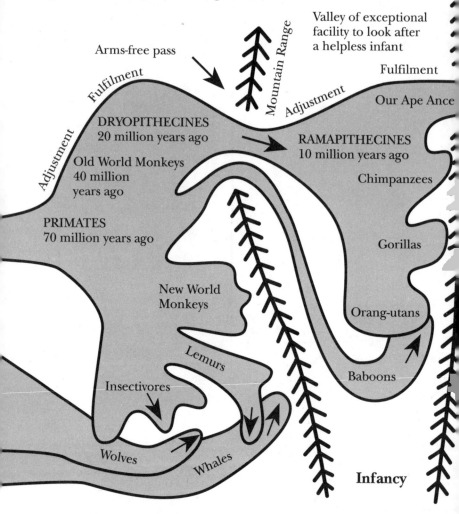

Note: *Adjustment* refers to the stage in Development when adjustments occur in response to an opportunity to develop. Major changes to organisms take place in this stage.

   Fulfilment refers to the fulfilment of the above adjustments. In this stage minor differentiations of a species can occur as different niches in the new realm and/or situation are occupied.

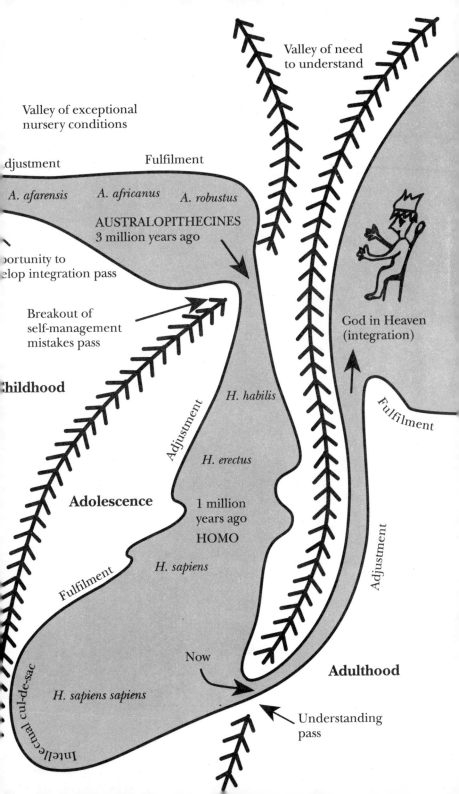

Valley of need
to understand

Valley of exceptional
nursery conditions

djustment   Fulfilment

A. afarensis   *A. africanus*   *A. robustus*

AUSTRALOPITHECINES
3 million years ago

ortunity to
elop integration pass

Breakout of
self-management
mistakes pass

God in Heaven
(integration)

hildhood

Adjustment

H. habilis

H. erectus

Adolescence

1 million
years ago
HOMO

Fulfilment

H. sapiens

Fulfilment

Adjustment

Now

Adulthood

Intellectual cul-de-sac

H. sapiens sapiens

Understanding
pass

Fig. 5: Primate Life Spans

The further a species can prolong infancy the more thoroughly the infants can be trained in integrativeness or be 'Love-Indoctrinated' resulting in more integrated adults.

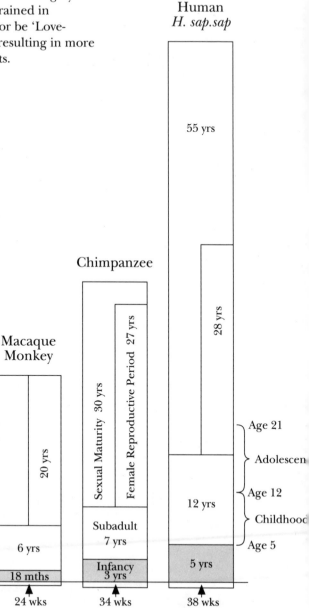

Human
H. sap.sap

55 yrs

28 yrs

Chimpanzee

Sexual Maturity 30 yrs

Female Reproductive Period 27 yrs

Macaque Monkey

20 yrs

Lemur

16 yrs

Subadult 7 yrs

6 yrs

Age 21

Adolescen

Age 12

Childhood

12 yrs

Age 5

Infancy 3 yrs

5 yrs

18 mths

18 mths

6 mths   18 wks
Gestation

24 wks

34 wks

38 wks

Illustrated Summary of the Development and Resolution of Upset

UNDERSTANDING the reason for our upset stops its growth and allows it to subside. With this resolution comes a paradigm shift, or monumental change in our way of thinking. Humanity steps out of an insecure 2-million-year adolescence into a secure adulthood.

[Judgement Day is] **Not the day of judgement but the day of understanding.**
Unnamed Turkish poet, mentioned in *National Geographic*, November, 1987.

The shift will be a time of <u>liberation</u>, <u>compassion</u> and <u>adjustment</u>. It will also bring <u>exposure</u> to and <u>confrontation</u> with the reality of our upset. There will be shock — the real 'future shock' — in adjustment but this massive transition can be managed.

'Future shock' . . . [is] the shattering stress and disorientation that we induce in individuals by subjecting them to too much change in too short a time.
Alvin Toffler, *Future Shock*, 1970.

The final section, *Adjusting to The Truth*, outlines the form of the new world in its transitional years.

Stage 1
The Origin of Upset

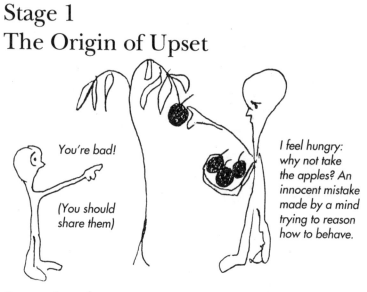

You're bad!

(You should share them)

I feel hungry: why not take the apples? An innocent mistake made by a mind trying to reason how to behave.

Our conscience, the expression of our original instinctive self or soul

Our mistake-prone mind or intellect

Our teenage years (the individual)
Sobered Adolescentman — *Homo habilis* (the species)

Our original instinctive self or soul (the expression of which is our conscience), perfectly orientated to integrative behaviour, recognises what is integrative ('good') and what is divisive ('bad'). However, it has no understanding of why integrativeness is meaningful and no appreciation of, or sympathy for, the intellect's need to understand.

When large-brained Adolescentman *Homo* emerged some 2 million years ago, he began experimenting in self-management. Whenever this had a divisive outcome (through a mistake in understanding), his conscience criticised him.

But while we needed guidance from our conscience, we didn't deserve its criticism. We were not bad as it implied we were. Tragically, we couldn't explain this.

136

Stage 2
Mildly Upset

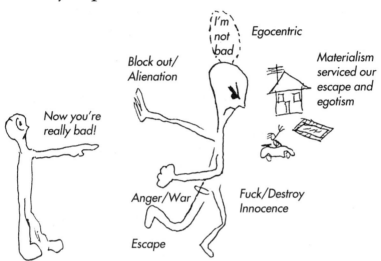

Our Twenties (the individual)
Adventurous Adolescentman — *Homo erectus* (the species)

Until we knew why we couldn't help making mistakes, we couldn't explain and thus defend them against our instinctive self's criticism. Consequently, we were forced to live in a state of insecurity and became upset. We resented the unfair criticism and tried to maintain our dignity or sense of goodness through demonstrations of our worth, and tried to block out and escape the criticism. We became angry, egotistical and alienated.

These expressions of our upset made us appear even more divisive, attracting further criticism from our conscience, compounding our insecurity. The more we searched for the understanding that would relieve our upset the worse we made our situation and the more our upset intensified.

Humphrey Bogart storms the terror-swept goldlands — a new high in high adventure! The nearer they get to their treasure, the farther they get from the law! And the more they yearn for their women's arms, the fiercer they lust for the gold that cursed them all!
From a poster for the film *Treasure of Sierra Madre*, 1948.

The threat of ignorance that emerged 2 million years ago was a threat against our species; if it prevailed we would never find understanding and never fulfil our responsibility to master intelligence. This battle to overthrow ignorance was so important it became our species' priority. When men in their role as group protectors went out to meet the threat, our species changed from matriarchal (female role dominated) to patriarchal (male role dominated).

Tragically, the task of championing the intellect made men angry, egocentric and alienated, or upset. Unable to defend their loss of innocence, men began to resent and attack innocence because it seemed to criticise their lack of innocence. The first victim was nature. Men began to 'hunt' (kill) animals because their innocence criticised them, albeit unwittingly.

The next victims were women. The rapidly compounding upset in men attracted criticism from women who were unaware of its cause. (Unable to explain their upset men couldn't even admit it — a failed attempt to explain it would only be misinterpreted by women as an admission of badness.) In retaliation against their criticism men attacked women. Since women reproduced the species, men couldn't destroy them as they did the animals. Instead they violated women's innocence or 'honour' by rape; they invented 'sex', as in 'fucking' or destroying, as distinct from the act of procreation. What was being fucked or destroyed was women's innocence. In this way women's innocence was repressed and they came to share men's level of upset. On a nobler level sex became an act of love. When all the world disowned men for their unavoidable divisiveness women stayed with them, bringing them the only warmth, comfort and support they would know.

The Lord God said, 'It is not good for the man to be alone. I will make a helper suitable for him' . . . Then the Lord God made a woman . . . and he brought her to the man.
The Bible, Genesis 2:18,22.

For copyright reasons I am unable to reproduce the lyrics of Bruce Springsteen's songs *Cover Me* and *I'm on Fire* from the album *Born in the USA* (1984), paraphrased in part as follows:
I won't go out there again because the whole world just wants to 'score'. I've seen enough of it and I don't want to see any more. I'm looking for a lover to come in and cover me.

Sometimes it seems as if someone's taken a blunt knife and cut a deep valley through my soul. I wake up at night soaked in sweat, with the feeling that a train's running through my head. I'm burning up and you're the only one who can quench my desire.

Sex and the relationship between men and women are subjects that deserve elaboration. Given that the whole world was an innocent friend of our soul but not of our apparently corrupt mind, it can be appreciated that men needed extremely powerful egos to defy ignorance and champion the intellect. Women, not responsible for the fight against ignorance, and so not partaking in the battle itself, did not and could not be expected to understand the battle. They could understand the search for the truth but not the battle involved.

Shirley MacLaine can't find a man to love. The 48-year-old actress . . . [said she] longs for a 'close and warm relationship' but hasn't met a suitable partner. 'Most men I meet seem to be too involved in trying to be successful or making a lot of money,'

she said. 'I feel sorry for all of them. Men have been so brainwashed into thinking they have to be so outrageously successful — to be winners — that life is very difficult for them. And it's terribly destructive, as far as I am concerned, when you are trying to get a serious relationship going.'

The *Daily Mirror* (a Sydney newspaper), December 14, 1982.

The only alternative to oppression was that men explain themselves to women but that was not possible. Men could not admit their inconsistency with integrative meaning until they could defend it.

One of the reasons that men have been so quiet for the past two decades, as the feminist movement has blossomed, is that we do not have the vocabulary or the concept to defend ourselves as men. We do not know how to define the virtues of being male, but virtues there are.

Asa Baber, *Playboy*, July, 1983.

(Feminists did not free themselves just because men stayed quiet as this quote suggests. The more men fought to defeat ignorance and protect the group [humanity] the more upset they became and the more they appeared to make the situation worse. The harder they tried to protect us the more they seemed to expose us to danger! In the end they became completely ineffective or inoperable, paralysed by this paradox. At this point women had to take over the day-to-day running of affairs as well as trying to nurture a new generation of soundness. Women, not oppressed by the overwhelming responsibility and extreme frustration that men felt, could remain effective. As well, when men crumpled women *had* to take over or the family, group or community involved would perish. A return to matriarchy, such as we have recently been seeing on earth, was a sign that men in general had become completely exhausted. However, it was not total matriarchy, because men could not afford to stand aside completely. They still had to stay in control of the fundamental battle and remain vigilant against the threat of ignorance. While some elements in the recent Feminist Movement seized the opportunity to take revenge against men's oppression, the movement in general was most necessary and valid.)

Women with infants digging roots.

Telling the Hunt.

These photographs from Lee and DeVore's *Kalahari Hunter-Gatherers* (1976), illustrate what is being said about the different situations of men and women. *Telling the Hunt* shows the men attentively hearing about — and presumably sharing in — the success of the hunt. We can imagine the hunter recounting how he relentlessly pursued his prey, methodically stalking and finally vanquishing it, to the cheers of his audience. You can sense the oppression men feel from the world at large; in a world that condemns them they are finding support in each other's company.

While the men are thus preoccupied satisfying their egos we find the women in the previous photo gathering food and looking after the children. While it was men's responsibility as group protectors to defy ignorance and champion the intellect or ego, women's conscious mind and thus ego also needed to be satisfied. (In fact if men had not existed women eventually would have had to take up the task of fighting ignorance, at which point they would have become as ego-embattled as men. Thankfully men did exist, allowing women to stay relatively innocent of the battle and free to nurture. It was a fortunate arrangement because fighting and loving are opposite qualities that would not easily coexist in the one person.) While women are not as ego-embattled as men they are necessarily so to a degree: they too have to try to understand the world and justify themselves against the criticism of that search for understanding. However, with men preoccupied with the main battle they had to fit in their search for, and fulfilment of their conscious thinking selves around their food gathering and nurturing. The one way this was possible was through talking to each other. They couldn't act out experiments in self-management but they could think and talk their ideas through, justifying and measuring themselves in words, and we can imagine that happening in this photo. They are undoubtedly keeping close together so they can talk.

What makes these photos such good illustrations is that they are of the relatively innocent Bushmen. While aboriginal peoples are necessarily more innocent than races more advanced (down the exhaustion curve) they are still members of the highly exhausted Sophisticatedman, *Homo sapiens sapiens* as these pictures confirm. The basic adaptations humans made to the human condition are clearly well established in the Bushmen. These could as easily be photos of businessmen discussing a takeover and women shopping.

We can see now how women's innocence fell victim to men's upset. Throughout the battle to find understanding women were being forced to suffer the destruction of their soul, their innocence, while at the same time their trappings of innocence were being cultivated. Originally (see page 90), cute, childlike features (domed forehead, snub nose and large eyes) were considered 'beautiful', and were favoured because they were the hallmarks of innocence and indicated a potentially integrative person. When ignorant innocence became a threat, men sought such 'beauty', such signs of innocence, for sexual destruction. We evasively described such looks as 'attractive' to avoid saying that what was being attracted was destruction, through sex, of women's innocence. Because all other forms of innocence were being destroyed, this image of innocence — 'the beauty of woman' — was the only form of innocence to be cultivated during humanity's adolescence. Women's beauty became men's only equivalent for, and measure of, the beauty of their lost pure world.

Sex is life.
Graffiti on a granite boulder at Meekatharra in Western Australia.

Women are all we [men] **know of paradise on earth.**
(Source unknown).

. . . we lose our soul, of which woman is the immemorial image.
Laurens van der Post, *The Heart of the Hunter*, 1961.

Woman stands before [man] **as the lure and symbol of the world.**
Pierre Teilhard de Chardin, *Je m'explique*, 1966 (published in English as *Let Me Explain*, 1970).

It was little wonder men fell in love with women. The 'mystery of women' was that it was only the physical image or object of innocence that men were falling in love with. The illusion was that women were psychologically as well as physically innocent. For their part, women were able to fall in love with the dream of their own 'perfection' — of their being truly innocent. Men and women *fell* in love. We abandoned the reality in favour of the dream. It was the one time in our life when we could romance — when we could be transported to 'how it could be' — to heaven.

It's timely to explain here why different cultures have had slightly different conceptions of beauty in women. Essentially, men are 'attracted' by innocent looks, which are youthful neotenous features. As the saying goes 'blondes have more fun' — that is, blondes are considered more attractive in Caucasian cultures because many young Caucasians have blonde hair, a sign of youth/innocence. Long, healthy hair is also associated with youth, which is why men find women's long hair attractive. In general, any feature unique to women will be attractive and signal a sex object to men, hence the desirability of breasts, shapely hips and narrow waist. The different cultural definitions of beauty can be explained in terms of what signals innocence. In times when few could eat or live well, fat women were considered beautiful because fat people were usually better fed and nurtured, indicating that they had been well cared for and were thus more innocent.

The destruction of women's souls and the cultivation of their image of beauty has been going on for 2 million years. Lust and the hope of falling in love have assumed such importance that many people, such as psychoanalyst Sigmund Freud, were misled into believing sex ruled our lives. However, underlying all this, the battle to find understanding always remained. Men and women became highly adapted to their roles. While men's magazines are full of competitive battleground sport and business, women's magazines are full of ways to become more 'attractive'.

One of the best examples of being misled into believing sex rules our lives is found in the story of the Garden of Eden, where Eve is blamed for tempting Adam to take the apple from the tree of knowledge. The truth is that men shouldered the responsibility of searching for knowledge, became upset by the criticism from ignorance and innocence and retaliated against all innocence, including destroying or fucking innocent women. Women were the victims not the cause of upset in men, but lust became such a strong force in us, we were misled into believing it seduced us into behaving in an upset way.

So 'attractive' did the object of innocence become that we eventually had to conceal it. Clothes were not developed for warmth as children are taught at school, but to dampen lust. The relatively innocent Bushmen go about almost naked most of the time, and often, naked. Once we became extremely upset, even the sight of a woman's ankle or face became dangerously exciting to men, which is why in some societies women are completely draped.

> **Then the eyes of both of them were opened, and they realised that they were naked; so they sewed fig leaves together and made coverings for themselves.**
> *The Bible*, Genesis 3:7.

The convention of marriage was invented as one way of containing the spread of exhaustion. By confining sex to a life-long relationship, the souls of the couple could gradually make contact and be together in spite of the sexual destruction involved in their relationship.

> **. . . a man will leave his father and mother and be united to his wife, and the two will become one flesh. So they are no longer two, but one.**
> *The Bible*, Mark 10:7,8.

Brief relationships kept souls repressed and spread soul repression. Ideally, if we wanted to free our soul from the hurt sex caused it, we needed to be celibate.

> **. . . others have renounced marriage because of the kingdom of heaven.**
> *The Bible*, Matt. 19:12.

In every generation, individual women had a very brief life in innocence before being soul-destroyed through sex. They then had to try to nurture a new generation, all the time trying to conceal the destruction that was all around and within them. Mothers tried to hide their alienation from their children, but the fact is if a mother knew about reality/ upset her children would know about it and would psychologically adapt to it. Alienation was invisible to those alienated, but to the innocent — and children are born innocent — it was clearly visible. For example Christ's mother Mary must have been innocent because we know he was. Since women become upset through sex, Mary must have had virtually no exposure to sex. The symbol for women's innocence/purity is virginity hence the description of Christ's mother as the Virgin Mary.

Women have had to inspire love when they were no longer loved or innocent, 'keep the ship afloat' when men crumpled and attempt to nurture a new generation; all the while dominated by men who couldn't explain why they were dominating, what they were actually doing or why they were so upset and angry! This was an altogether impossible task, yet women have done it for 2 million years. It was because of women's phenomenally courageous support that men, when civilised, were chivalrous and deferential towards them. Men had an impossible fight on their hands, but at least they had the advantage of understanding the battle.

With men defying and repressing their souls, women became the representative of the soul in the partnership between men and women. As well, because of men's unexplained oppression of women and the world of the soul, and women's inability to understand it, women were forced to rely on and trust their soul more than their ability to understand. As a consequence, they are more intuitive or dependent on their soul's guidance than men. A common euphemism is 'women feel and men think'.

The sword was, he would suggest, one of the earliest images accessible to us of the light in man; his inborn weapon for conquering ignorance and darkness without. This, for him, was the meaning of the angel mounted with a flaming sword over the entrance to the Garden of an enchanted childhood to which there could be no return. He hoped he had said enough to give us some idea of what the image of the sword meant to him? But it was infinitely more than he could possibly say about the doll. The doll needed a woman not a man to speak for it, not because the image of the sword was superior to the image of the doll. It was, he believed, as old and went as deep into life. But it was singularly in women's keeping, entrusted to their own especial care, and unfortunately between a woman's and man's awareness there seemed to have been always a tremendous gulf. Hitherto woman's awareness of her especial values had not been encouraged by the world. Life had been lived predominantly on the male values. To revert to his basic image it had been dominated by the awareness of the sword. The other, the doll, had had to submit and to protect its own special values by blind instinct and intuition.

Laurens van der Post, *The Seed and the Sower*, 1963.

(Note that these truths about the different roles of men and women are further examples of truths that we had to evade until we found the defence for our divisive nature. While men and women are different, sexist notions of men being 'evil' or of women being irrelevant have no credibility. To avoid prejudice we maintained that what difference there was between men and women was not profound, but simply a product of cultural conditioning — of girls being given dolls and boys swords as infants for example. In fact, as Sir Laurens van der Post agrees, our cultural differences are the product of very real differences between the sexes.)

The development of women's exhaustion was tied to men's. Women had to try to 'sexually comfort' men but also preserve as much real innocence in themselves as possible for the nurturing of the next generation. Their situation, like men's, worsened at an ever-increasing rate. The more women 'comforted' men, the less innocence they retained and the more the next generation suffered and needed 'comforting', etc. If humanity's battle had continued for a few thousand years more all women would have become like Marilyn Monroe, complete sacrifices to men. At this point men would have destroyed themselves and the species because there would have been no soundness left in women to love/nurture new generations.

> [when talking of men persuading women to have sex Olive Schreiner said (through her female character) men say] . . . **'Go on; but when you** [men] **have made women what you wish, and her children inherit her culture, you will defeat yourself. Man will gradually become extinct . . .' Fools!**
> *The Story of an African Farm*, 1883.

Sex killed innocence. During humanity's adolescence that was what sex was all about, although it was also one of the greatest distractions and releases of frustration and, on a higher level, an expression of sympathy, compassion and support — an act of love.

> **Touched by her concern for her honour, in his imagination he would have liked to tell her that he could kneel down before her as a sign of how he respected her and beg her forgiveness for what men had taken so blindly and wilfully from women all the thousand and one years now vanishing so swiftly behind them. But all he hastened to say was: 'I would have to be a poet and not a soldier to tell you all that I think and feel about you. I can only say that you are all I imagined a good woman to be. You make me feel inadequate and very humble. . . . Please know that I understand you have turned to me not for yourself, not for me, but on behalf of life. When all reason and the world together seem to proclaim the end of life as we have known it,** [this conversation is taking place on the eve of a World War II battle] **I know you are asking me to renew with you our pact of faith with life in the only way possible to us.'**
> Laurens van der Post, *The Seed and the Sower*, 1963.

The paradox was that having destroyed innocence, men would end up wanting to rediscover it. The truth was that men were having to repress and 'hurt the ones they loved'.

I thought finally that of all the nostalgias that haunt the human heart the greatest of them all, for me, is an everlasting longing to bring what is youngest home to what is oldest, in us all.
Laurens van der Post, *The Lost World of the Kalahari*, 1958.

While women's oppression has been extreme so has men's. Men have yearned for freedom from their oppressor, ignorance, as much as women have yearned for freedom from their oppressors, men.

' . . . if I might but be one of those born in the future; then, perhaps, to be born a woman will not be to be born branded. . . . It is for love's sake yet more than for any other that we [women] look for that new time. . . . Then when that time comes . . . when love is no more bought or sold, when it is not a means of making bread, when each woman's life is filled with earnest, independent labour, then love will come to her, a strange sudden sweetness breaking in upon her earnest work; not sought for, but found.'
Olive Schreiner, *The Story of an African Farm*, 1883.

It should be pointed out that our destruction and corruption of innocence has been going on at all levels. We even destroyed our own innocent soul by repressing it. All forms of innocence unfairly criticised us, so all forms of innocence were targets for our attack. Sunglasses aren't always worn to shade the eyes from the sun. Often they were worn to alienate ourselves from the natural world that was alienating us. They were an attack on the innocence of daylight.

Another adaptation to life under the human condition was the development of materialism. In a world ignorant of our true goodness, we sought material glorification. While spiritual relief (understanding) had still to be found material relief was all that was available.

(In our highly embellished material world order, as an aspect of integrative meaning or God, critisiced us for our apparant disorder and lack of integrity; in retaliation we often deliberately distorted order and cleanliness of elements and colour in our designs and decorations. In our clothing and home decoration for instance, we preferred the come and go of fashion or change to constant, unwavering and unforgiving functional integrity. In fact the defensive word used against purity in design was 'boring'.)

Stage 3
Extremely Upset

Lots of materialism now. While spiritual relief (understanding) had still to be found, only material relief was available.

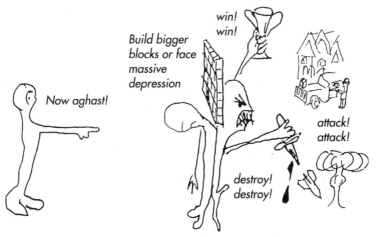

Our thirties (the individual)
Angry Adolescentman — *Homo sapiens* (the species)

Until we knew why we made mistakes and could defend ourselves against criticism from our conscience, our insecurity of self and its resulting upset only increased. Eventually we became extremely upset.

Stage 4
Abandon the Battle and
Transcend our Upset

1. Civility's restraint of upset
2. Religion's born-again idealism
3. Marx's enforced idealism
4. New Age & Environmentalist
   non self-confronting idealism
5. Escape through drugs
6. Return to superstitious answers
7. Stop thinking

Our Forties (the individual)
Sophisticated Adolescentman
— *Homo sapiens sapiens* (the species)

The price of searching for knowledge was that we became
upset. We became angry, egocentric and alienated.

I should explain that our alienation developed on two
fronts; the separation of our intellect from the truth, which
was our neurosis or oppression of our intellect, and the
separation of our intellect from our soul, which was our
psychosis or repression of our soul. We stopped thinking
and repressed our sensitive instinctive self. We died in both
intellect and soul.

To explain our neurosis, we stopped thinking because it confronted us with the integrative ideals that we couldn't mentally defend ourselves against. Deep, truthful, profound thinking led our mind into depression. To avoid the depression our thinking became increasingly superficial or escapist.

Thinking can get you into terrible downwards spirals of doubt.
Rod Quantock, *Sayings Of The Week*, *Sydney Morning Herald*, July 5, 1986.

To explain our psychosis, we repressed our soul because it reminded us of the unjustly condemning integrative ideals.

Alienation was a wretched state. It was a life of darkness, a living hell.

. . . banished . . . from the Garden of Eden . . . hidden from your presence . . . a restless wanderer on the earth.
The Bible, Genesis 3:23,4:14,12.

Genghis Khan lived in extreme upset. He lived out his anger, egocentricity and alienation to the full. Every day he satisfied his anger with bloodletting, his egocentricity by domination of others and his mind or spirit by blocking out any feeling of guilt. There would have been no peace in his life; no freedom for his troubled spirit.

So what could we do when we'd become unbearably upset but still hadn't found the answers that would free us from criticism? Only abandon the search and try to return to the world of our soul and its idealism.

But first we civilised our upset. We restrained and contained the expressions of our upset rather than stop the search. However, as upset increased, we eventually *had* to abandon the search.

The first form of abandonment was religion, through which we could be 'born (to the ideals) again'. The beauty of religion was that while we personally had abandoned the battle, it continued through the innocence of the prophet our religion was founded around. The task of confronting ignorance continued through the support we gave his soundness. Carl Jung was fond of saying **the Christian symbol is a living being that carries the seeds of further development within itself** and **in Christianity the voice of God can still be heard** (*The Undiscovered Self: Present and Future*, 1961).

But by retaining a presence of innocence, religion reminded us of our loss of it, accentuating our sense of guilt. The more upset we became the more intolerable any implication of guilt or badness became, since we knew the greater truth was that we were not guilty or bad. Therefore the greater our exhaustion or upset the more we needed forms of idealism to support that didn't remind us of our lack of innocence. Less self-confronting, more guilt-free forms were needed. But this meant less acknowledgement of the truth of our lack of soundness. The trend was towards greater sophistication in evasion or superficiality or alienation.

Following religion came socialism and communism, which imposed social or communal (integrative) ideals and oppressed freedom to search for understanding, but unlike religion didn't maintain recognition of the world of soundness or soul. There was no element of faith that there were some who could confront the truth. The attraction was that we could support and have the ideals without confronting self, which means socialism and communism were more sophisticated than religion in their degree of evasion or superficiality.

Most recently, the New Age Movement developed. It advocated that we adopt positive thoughts about ourselves — learn to rise above or 'transcend' our real but insecure and upset selves. In socialism, while there was no confrontation with our upset selves there was still acknowledgement of the cooperative or integrative ideals. In the 'human potential' stressing New Age Movement all the emphasis was on self-affirmation. The depressing integrative ideals and our upset soulless side were overlooked in favour of just holding onto the greater truth that we are good. Complete superficiality had arrived. All negatives were removed. The New Age Movement signalled the arrival of total alienation (self-deception or dishonesty). It was the ultimate sophistication in evasion. The truth is it was leading us not to a new age of freedom from upset but to the ultimate alienated state!

The capacity for transcendence brings with it a liability to alienation, and the wish to escape this condition . . . can lead to even greater absurdity.
Thomas Nagel, *The View From Nowhere*, 1986.

Abandoning the upsetting search for understanding and the self-confrontation involved could contain and temporarily heal our upset, but it could not and did not win the battle. Our real freedom from upset depended on finding the understanding of our greater goodness that would end our insecurity. To be genuinely free of upset we had to confront and overthrow ignorance. Ultimately we had to succeed at thinking and vindicate the intellect, not throw it away. The purpose of a conscious mind was to understand, and the responsibility of having a conscious mind was to achieve that understanding.

. . . I am infinitely saddened to find myself suddenly surrounded in the west by a sense of terrible loss of nerve, a retreat from knowledge into — into what? Into Zen Buddhism; into falsely profound questions about, Are we not really just animals at bottom; into extra-sensory perception and mystery. They do not lie along the line of what we are now able to know if we devote ourselves to it: an understanding of man himself. We are nature's unique experiment to make the rational intelligence prove itself sounder than the reflex. Knowledge is our destiny. Self-knowledge, at last bringing together the experience of the arts and the explanations of science, waits ahead of us.
Jacob Bronowski, *The Ascent of Man*, 1973.

To be truly free we had to confront our condition. False prophets taught people ways to escape their condition. They led them away from the battle. Real prophets dug up the truth.

. . . many false prophets will appear and deceive many people. *The Bible*, Matt. 24:11.

Far from it ever having been wrong to abandon the battle and rest when we became exhausted, rest and recuperation have been vital. (With regard to recuperation — in religion, prayer, chanting and meditation were used to suppress the intellect and allow the soul to the surface; in socialism and communism work was the therapy, the means for shutting out the worried mind. In the New Age Movement our re-pressed sensitivity found emotional release through hugging each other, channelling and identification with rainbows, stars, dolphins, crystals and pyramids. Again we see the trend towards superficiality.) What was false and dangerously mis-leading was to claim that escaping the battle to find under-standing was the way to win it, and the path to follow. (Tragically there was no alternative to self-deception/false-ness. Admitting that we were escaping would make us feel that we were bad when we knew that we weren't. Until the arrival of the full truth that defended our exhaustion, self-deception and its superficiality were unavoidable.)

The danger was that, as exhaustion or upset spread, the need for fundamentalist obedience to the ideals, oppression of the upsetting search for understanding and transcend-ence of self with its superficiality might expand to take over the world, ending all search for understanding.

If you want a picture of the future, imagine a boot stamping on a human face [freedom] **— for ever**. George Orwell, *1984*, published in 1949.

If everyone escaped there would be no-one left to confront the problem of ignorance, in which case we would never achieve liberation from the human condition. Escape led to oblivion.

He [the self-deception that accompanies superficiality] **will invade the kingdom** [of honesty] **when its people feel secure** [when superficiality becomes popular enough], **and he will seize it** [the kingdom of honesty] **through intrigue . . . Then they** [those pushing self-deception] **will set up the abomination that causes desolation** [the superficiality that leads to oblivion]. **With flattery he will corrupt those who have violated the covenant** [self-deluding superficiality will seduce the exhausted], **but the people who know their God will firmly resist him** [the less exhausted will not be deceived].
The Bible, Dan. 11:21,31,32.

So when you see the 'abomination that causes desolation' (spoken of through the prophet Daniel) standing where it does not belong [claiming to know the way to the new age] let the reader understand . . . For then there will be great distress [mindless superficiality and its consequences], unequalled from the beginning of the world until now — and never to be equalled again. If those days had not been cut short [by the arrival of the truth], no-one would survive.
The Bible, Matt. 24 and Mark 13.

The trend has been towards greater self-deception or evasion (alienation). Figures 2 and 3 (pages 130 – 131) show that while *Homo sapiens sapiens* had become highly intelligent (Fig.2) and with that intelligence gained understanding or become wise (*sapiens* means wise) he was at the same time becoming more sophisticated in evasion or alienated or false or departed from the integrative truths and ideals (Fig.3). *Homo* wise wise should have also been called *Homo* false false. Emphasising only our wisdom was an evasion of our falseness — an example of our self-deception or sophistication in evasion or alienation.

The hope was that we would find understanding before we became so committed to escape that there would be no-one left trying to confront and reveal the truth about us.

> **. . . the one primary and elemental approach to the problem** [that the world faces] **is through the being of man. Unfortunately it is an increasingly lonely way, trodden more and more not by masses but by solitary individuals . . .** [only these few] **sustain** [man's] **urge to seek an answer to the riddle of life . . .** Laurens van der Post, *The Dark Eye in Africa,* 1955.

Now that the truth is revealed the remaining task is to have everyone confront it. Humanity's 'brave new world' is the journey into self to liberate our repressed psyche or soul. The final section, *Adjusting to The Truth,* explains how facing the truth about ourselves can be managed.

Developing Answers

Stage A
The Search

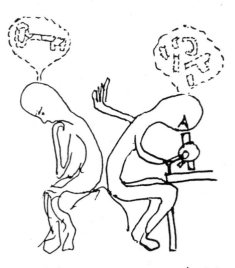

*Unevasive holistic
subjective introspection*

*Evasive mechanistic
objective science*

To refute the implication that we were bad, we needed to explain why we were not. We had to learn why we were the way we were. We had to find our identity. Searching for our identity or meaning or context or worth was the concern of humanity's adolescence.

We had to explain why we made mistakes. We had to find the good reason for our apparent badness and, in so doing, end our insecurity and the upset it caused us. In religious terms we needed to discover the origin of sin.

To find this key understanding that would liberate us, we first had to learn the mechanisms of our world. In particular we had to know that nerves are insightful and genes are not. Finding these mechanisms was the role of science.

While looking for this knowledge, science had to evade integrative ideals because they unjustly criticised our divisive upset state. We had to be free from the integrative ideals to search for understanding of those ideals. This also meant science had to avoid subjective introspection (soul guidance) and depend on objective research. We had to be free from the condemning integrative ideals that our conscience insisted on.

However, while science had to be evasive and objective, we also needed unevasive subjectivity to complete the journey. Unevasive thought was needed to assemble the full truth from science's hard-won but evasively presented insights. The full truth that explains us couldn't be reached via evasion. For example while denying the existence of integrative meaning we were in no position to recognise that our soul was our genetic orientation to integration. We couldn't begin to look into the human condition while being evasive. After all, our evasion or insecurity is the human condition. Alienation couldn't investigate alienation. In the end sound, secure, unevasive soul guidance was needed to reveal the truth about ourselves.

Only those free of upset could look into the human condition. Innocence would deliver the key to our freedom from upset. Exceptional innocence — not exceptional cleverness — would lead us home to integration or peace. 'Think tanks' of exceptionally clever people like Nobel Prize winners and academics were not going to solve the world's problems. Their contribution was finding the mechanisms while evading any unjustly condemning partial truths. They were the custodians of the evasion that eventually monopolised the world. Clever, esoteric intellectualism was a cul-de-sac in Development. The answers would come from back down the path of our development towards non-clever, innocent soundness. It wasn't 'brain-storming' but 'soul-storming' that would free us.

**The wolf will live with the lamb,
the leopard will lie down with the goat,
the calf and the lion and the yearling together;
and a little child will lead them.**

. . .

for the earth will be full of the knowledge of the Lord.
The Bible, Isaiah 11:6,9.

Whatever happens, I shall be there in the end, for I, child that I am, am mother of your future self.
Laurens van der Post, *Jung and the Story of Our Time,* 1976.

One of the most moving aspects of life is how long the deepest memories stay with us. It is as if individual memory is enclosed in a greater which even in the night of our forgetfulness stands like an angel with folded wings ready, at the moment of acknowledged need, to guide us back to the lost spoor of our meanings.
Laurens van der Post, *The Lost World of the Kalahari,* 1958.

There is the biblical story of David slaying Goliath, and in what is probably Australia's most famous poem, Banjo Paterson's *The Man From Snowy River* (1895), a boy brought in the wild horses after the men had failed, and in Hans Christian Andersen's story *The Emperor's New Clothes* (1837) it is a child who breaks the spell of deception and discloses the truth.

Evasive, mechanistic, objective science and unevasive, holistic, subjective introspection had separate roles to play in liberating humanity from criticism, insecurity and guilt.

Science without religion is lame, religion without science is blind.
Albert Einstein, *Out of My Later Years*, 1950.

Perhaps this life of ours, which begins as a quest of the child for the man, and ends as a journey by the man to rediscover the child, needs a clear image of some child-man, like the Bushman, wherein the two are firmly and lovingly joined in order that our confused hearts may stay at the centre of their brief round of departure and return.
Laurens van der Post, *The Lost World of the Kalahari*, 1958.

Sir Laurens van der Post's unevasive books about the Bushmen have been the inspiration for my own writing. The Bushman's comparitive innocence sheltered my soul, giving it companionship and confirmation. Without the Bushmen my soul would have been too alone to stand up to our denial of our souls' world and all the truths that reside there. That is why this book is dedicated to Sir Laurens van der Post.

As the quotes in this book reveal, all I have been able to add to the perception/soundness of Jesus Christ and Sir Laurens van der Post is the biological reason for the repression of our soul. Through all the evasion in recorded history a thread of strong, unevasive thought has been held onto and developed. The exceptionally unresigned and unevasive thinkers or prophets of modern times that I have become aware of are Sir Laurens van der Post, Pierre Teilhard de Chardin, Arthur Koestler, Eugène Marais, William Blake, Jean-Jacques Rousseau and Antoine de Saint-Exupéry.

From the days of John the Baptist until now, the kingdom of heaven has been forcefully advancing, and forceful men lay hold of it. For all the Prophets and the Law prophesied until John.
The Bible, Matt. 11:12.

I stress again that science, and in fact humanity as a whole, is the real liberator of humanity from the human condition.

In many ways prophets only got in the way while we were searching for understanding because they confronted us with truths that depressed us and which we therefore had to evade. Exceptional innocence played an important but minuscule concluding role in our search for knowledge. In gridiron football the team as a whole (with one exception) does all the hard work gaining yardage down the field. Finally when the side gets within kicking distance of the goal posts a specialist kicker, who until then has played no part, is brought onto the field. While he — in his unsoiled attire — kicks the winning goal, the win clearly belongs to the exhausted players who did all the hard work.

The concepts presented in these pages are a synthesis (and reconciliation) of biology, physics, chemistry, philosophy, psychology and indeed all the scientific disciplines, and they were 'disciplines'. To be prepared to put aside the big questions of life's meaning and our inconsistency with it and go in search of an understanding of the details and mechanisms of the workings of our world required great restraint, perseverance and patience.

These answers were built on the wealth of detail won at great personal sacrifice by the warriors of all the scientific traditions. But science was only the institution humanity created to investigate the mechanisms of our world. Science itself depended on the supportive structure of civilisation for its continued existence. In truth every human who has ever lived has contributed to this breakthrough. It stands on the shoulders of all the weary battlers who fought the only way they could; every businessman, selfless or greedy; every nurse, compassionate or uncompromising; indeed every victim of paradox from the birth of the concept 'I'.

Since the emergence of our fully conscious mind over 2 million years ago, humanity has dreamed of, hoped and prayed for and, above all, worked unceasingly against immense odds towards this breakthrough.

Stage B
Reconciliation

New World   Old World

Objective research, assisted in the end by subjective intro-spection and supported by humanity as a whole has found the key to our freedom.

OUR SOUL AND INTELLECT ARE RECONCILED.

We can now return to integrativeness. The way forward now is back to our soul's world of soundness, naturalness, happiness and peace, and when we arrive this time we will be fully conscious and understanding. We have turned our heads for home. Our journey away through the wilderness and darkness is finally over.

The poles of life are reconciled: God (integrativeness) and (divisive) man, women and men, young and old, religion and science, holism and mechanism, emotion and reason, left wing and right wing, idealism and realism, yin and yang, Abel and Cain, blacks and whites, the selfless and the selfish, the natural and the artificial, the non-sexual and the sexual, the non-material and the material, the innocent and the born-again.

While Michelangelo's masterpiece (pictured on page 98) is titled *The Creation of Adam* it also became a symbol of our hope of reconcilation with God. The 'out-stretched fingers' of God and man have finally touched. Our estrangement has ended.

To quote the final words in Andrew Lloyd Webber's musical *The Phantom of the Opera* (1986); **it's over now, the music of the night**. Great optimism, relief, fellowship and excitement emerges now. Light comes into our darkness and will spread across the Earth until there is no darkness left. Our music will change. Our 'classical' music expressed the greater truth of our true divinity, the truth that stood above the terrible suffering the human condition inflicted on us. We could identify with its message but we couldn't relieve our condition. We waited in hope and faith for understanding. It *was* the music of the night. Now a new music will appear that expresses an extraordinary relief and happiness. A new world is coming for humans.

The age-old clenched fist symbol shows strength and defiance against any implication that we are bad. It can now hold aloft the key to liberation of the proof of our goodness. The psychological repair of humanity can begin.

Understanding that we are good allows our upset to subside. It gives us **the keys of the kingdom of heaven** (*The Bible*, Matt. 16:19). Understanding is compassion — it doesn't condone upset, it heals it.

Because these explanations reconcile intellect and soul they can be tested by either or both. That they satisfy both will prove their validity. They make so much sense that many of us will feel we are being brought out of a trance, and are so obvious that we will wonder why they weren't brought to light much sooner.

These explanations are grounded in first principle biology and are rational. No faith is required. This is the end of faith and belief and the beginning of knowing. The conclusions are logical. Freed from our dependence on evasion, we can reason them out for ourselves. The only way we can deny these explanations and their implications is by continuing to be evasive, and evasion in the presence of the truth won't succeed for long. Honesty and with it freedom for humanity is now only a matter of time.

> **. . . the Truth has to appear only once . . . for it to be impossible for anything ever to prevent it from spreading universally and setting everything ablaze.**
> Pierre Teilhard de Chardin, *Je m'explique*, 1966 (published in English as *Let Me Explain*, 1970).

Note also that these explanations are not discoveries, but revelations of things we have known; the many criticising partial truths that make up the full truth, such as integrative meaning and the significance of nurturing in our upbringing, were truths we knew but have evaded. To evade something we had first to know it. Nearly all the quotes in this book represent instances when the wall of our evasion or alienation or blindness was momentarily penetrated and the truth let through. We can find the truth amongst all our evasions now. In fact we can know exactly how alienated/evasive any writer was. Everything from the old world will be measured now in terms of its degree of alienation. The truth has been made clear. This is **the day** [predicted in *The Bible*] **when God will judge men's secrets** (Romans 2:16). This is the day of reckoning but it will be a kind time, not hurtful.

The problem had become one of revealing the truth, not of finding it. In the end it was our alienation that stood between us and freedom from the human condition. As explained earlier, unevasive or innocent thought was required to reveal these answers.

. . . you have hidden these things from the wise and learned, and revealed them to little children.
The Bible, Matt. 11:25.

The confidence or certainty of unevasive or holistic thinking will (mistakenly) seem arrogant to our evasive, mechanistic minds. If you are in a lighted room as it were, and someone asks where the door is, you can confidently say it <u>is</u> there. If you are being evasive on the other hand, if you are in a dark room and someone asks where the door is, you say I think it is <u>possibly</u> over there somewhere. Evasive thinking is blind but unevasive thinking isn't. Living with the truth, an unevasive mind knows when it is right and when it is wrong in its thinking. This is why unevasive or non-alienated people (such as the founders of the great religions) taught **as one who had authority, and not as their teachers of the law** (*The Bible,* Matt. 7:29).

Also, having to be evasive made the task of finding understandings extremely difficult. An unevasive mind, able to think straight/truthfully, can think so effectively that an evasive mind looking on would be astonished. It was asked about Christ, who was exceptionally sound or non-alienated, **How did this man get such learning without having studied?** (*The Bible,* John 7:15). Exceptionally unevasive thinkers or prophets could, for example, prophesy or know the future because they could think truthfully and thus effectively. Prophets were truth-tellers while most people had to evade the truth. Prophets were able to **delight in the fear of the Lord** (*The Bible,* Isaiah 11:3). Relatively innocent or free of upset, they weren't afraid of integrative meaning. They had never become resigned to reality (abandoned ideality and become evasive).

It won't be easy at first for the old, evasive world to adjust to the new, unevasive one; the evasive world will be incredulous and will try to impose its insecure view on the secure, unevasive one, receding slowly.

Adjusting to The Truth

The Activities of the New World

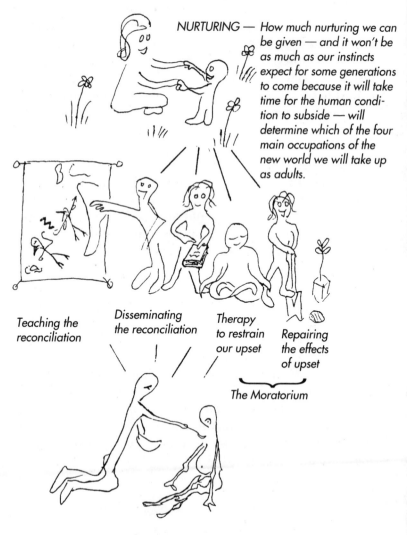

NURTURING — How much nurturing we can be given — and it won't be as much as our instincts expect for some generations to come because it will take time for the human condition to subside — will determine which of the four main occupations of the new world we will take up as adults.

Teaching the reconciliation

Disseminating the reconciliation

Therapy to restrain our upset

Repairing the effects of upset

The Moratorium

Now that psychological repair can begin, there are many adjustments to make. The most difficult will be confronting and accepting the extent of our upset and the great differences in upset between us.

When we realise this exposure, our first reaction will be to try to deny it. But this 'old world' evasive tactic won't work in the presence of full truth. We'll discover we can't 'old world' the 'new world'. Evasion has lost its power.

How are we to cope with exposure? How are we to adjust to the new world? The answer is, by choosing a way of participating in the new world that gives us time to adjust to the truth. There are four ways; <u>while equally important, they differ in the degree of confrontation with the truth that they require.</u> We each choose a way of participating that suits our level of upset. If we don't confront the truth personally we will still be evading it to a degree, but the important point is we won't be denying it as in evasive old world practices. In the old world we had to learn how to hide the truth. Now we have to learn how to live with the truth present. With the full truth of our goodness found this can be done, but it is an entirely new way of living for us.

<u>Essentially all we have to do now is support the unevasive truth</u> (hold the key aloft). <u>It is not necessary to confront it.</u> This is the way to cope with judgement day (exposure day — confrontation day — revelation day — truth day).

Of course in reality we will each participate in all the ways although to varying degrees. Each day will be a mix of the four activities that will vary from person to person according to our level of alienation. To begin with most of us will only be able to confront the truth (activity 1) a little, but as our alienation heals our strength will increase. While it may take a generation or two, eventually we will all absorb the truth fully and not be false any longer.

(1) Investigate and add to the unevasive understanding and teach it to others, especially the young. (It's far easier for young people to adopt unevasive understanding than for people who grew up in the old resigned/evasive paradigm.

Youth does not have to change a practised way of thinking and justifying itself. The young are the real beneficiaries of the truth. Those who have grown up in the old, evasive paradigm are the 'lost generation' that we have long intuitively known would result when the truth arrived. While they can fully support and participate in the new world they will have difficulty adopting its thinking. Physicist Max Planck succinctly described the difficulty adults have adopting new ideas when he said that science progresses funeral by funeral [see his *Scientific Autobiography*, 1948]. Often the only way adults, especially males, could adopt new ideas was by re-working them to appear as their own findings. This is a measure of how embattled our egos became. In any case, using new world unevasive understandings to bolster the ego instead of to abandon it would be self-defeating because it would only serve to bring the truth of our falseness to light. Further, when exposure arrives those without true vision will be found wanting in their ability to guide us in the new world and to add to the unevasive understandings. We will soon learn that self-deception is futile in the new world and that what works — and works incredibly well — is honesty.)

The difficulty of this way of participating where we investigate, add to and teach the unevasive truth is that it requires a great deal of confrontation with unevasive truth. Because of this it is the more innocent, or least alienated among us, who have little upset to confront, who will be most suited to this activity. Since the reality on Earth at present is that most people are extremely exhausted (in intellect and soul) very few will be able to participate in this way initially.

(2) Help disseminate and promote the unevasive understandings. This requires some indirect confrontation with the unevasive truth, but nowhere near as much as having to study and teach the understandings.

(3) and (4) (below) require little direct confrontation with the truth and are for the most upset or battle weary.

We must now call a moratorium on our upset. Until it is healed through understanding we have to restrain it and

170

contain its outward effect on the world. We must buy the time we need to confront and absorb the unevasive truth, to ensure that we have a world to live in when our upset does eventually subside.

(3) Restrain our upset. One way of doing this is to learn from the many techniques now available for quietening it and restoring the soul, such as meditation, prayer, yoga, chanting of mantras and communion with nature. (Nature revitalises our soul because our species grew up with it. We are instinctively adapted to it. It's our *natural* environment. The word 'natural' is our evasive code word for 'soul compliant'. The Environment Movement has rarely acknowledged the real value of nature, saying for example that we need the forests for new drugs and to maintain the climate. Nature's real importance to us is as our soul's companion.)

Another way of restraining our upset is to rise above or transcend it so it doesn't break out. What is required here is that we choose ways to transcend our upset that don't deny its existence.

Again it must be emphasised that while techniques for restraining our upset relieve and even temporarily heal it, they are not the final solution to upset. That depends on confronting it and dissolving it with understanding. Since confrontation takes time we have to rest and deny our upset to contain it in the interim. We need people who can teach resting and transcending upset. This work will suit those who find confronting the unevasive truth hurtful and who know what it is like to be exhausted.

We can see now that by promoting resting and transcendence the New Age Movement anticipated the time when we could go to work for the new world. We can also see that it was false to claim that resting from the battle and transcending its effects would bring about the new age as its title implied. Abandoning the battle was not going to win it. With the battle now won this self-deception is removed and New Age techniques for resting and transcending upset become most valuable.

(4) Contain and repair the expressions and effects of our upset. The Environment or Green or Conservation Movement anticipated this way of participating in the new world. The Environment Movement was token idealism dealing with the symptoms, not the cause of the problem, the human condition. With the human condition solved The Environment Movement also loses its falseness. It too can now take its proper place in the new world.

The maintenance and repair of our environment requires the least confrontation with self. (This has been its appeal. To quote *Time* magazine [December 31, 1990] **The environment became the last best cause, the ultimate guilt-free issue**.) This is clearly the way for the most exhausted or upset among us to participate in the new world.

We can see that these four new world activities are ordered according to our level of innocence. In the old world we had to evade the whole concept of innocence because of its unjust criticism of those who were no longer innocent. In the old world the innocent were never acknowledged. They were anonymous people, unwanted and alone. (To say we don't know who is innocent is untrue; to ignore, deny, repress and in the extreme crucify it, as we have done, we had to recognise it.)

Foxes have holes and birds of the air have nests, but the Son of Man has no place to lay his head.
The Bible, Matt. 8:20.

Samuel Taylor Coleridge described our **soul in agony**:
Alone, alone, all, all alone,
Alone on a wide wide sea!
The Rime of The Ancient Mariner, 1797.

Generally, the only identity afforded the innocent was as fools; it was unsafe (foolish) to present dangerous partial truths without the defence for our divisiveness. The real fools, however, in the sense of being the least sound or most alienated from the truth and thus the least able to think straight and effectively investigate the truth, were those who were sophisticated in the art of evasion, the so-called wise.

Do not deceive yourselves. If any one of you thinks he is wise by the standards of this age, he should become a 'fool' so that he may become wise. For the wisdom of this world is foolishness in God's sight. As it is written: 'He catches the wise in their craftiness'; and again, 'The Lord knows that the thoughts of the wise are futile'.
The Bible, I Corinthians 3:18,19,20.

When I saw [Shakespeare's] *Lear* here, I asked myself how it was possible that the unbearably tragic character of these fools had not been obvious long ago to everyone, including myself. The tragedy is not the sentimental one it is sometimes thought to be; it is this:

   There is a class of people in this world who have fallen into the lowest degree of humiliation, far below beggary, and who are deprived not only of all social consideration but also, in everybody's opinion, of the specific human dignity, reason itself — and these are the only people who, in fact, are able to tell the truth. All the others lie.

   In *Lear* it is striking. Even Kent and Cordelia attenuate, mitigate, soften, and veil the truth; and unless they are forced to choose between telling it and telling a downright lie, they manoeuvre to evade it.
Simone Weil, An Anthology, edited and introduced by Sian Miles, 1986.

The intellectual, the clever, the self-deceiving, the wealthy, the powerful, the aggressive and the egocentric dominated the old world. Now that our loss of innocence is defended, there's no reason to repress innocence and deny its soundness, or to be evasive.

The meek . . . will inherit the earth.
The Bible, Matt. 5:5.

. . . many who are first will be last.
The Bible, Matt. 19:30,20:16; Mark 10:31; Luke 13:30.

The order is rapidly fadin'
and the first one now will later be last
For the times they are a-changin
From Bob Dylan's song *The Times They Are A-Changin,* 1964.

The stone the builders rejected has become the capstone . . .
The Bible, Ps. 118:22; Matt. 21:42; Mark 12:10; Luke 20:17; Acts
4:11; 1 Pet. 2:7.

It is part of the great secret which Christ tried to pass on to us
when He spoke of the 'stone which the builders rejected'
becoming the cornerstone of the building to come. The corner-
stone of this new building of a war-less, non-racial world, too, I
believe, must be . . . those aspects of life which we have despised
and rejected for so long.
Laurens van der Post, *The Dark Eye in Africa,* 1955.

After 2 million years of having to repress innocence we
now bring it forward both in ourselves and in the world. We
let innocence live now. Resurrecting innocence is the new
direction for humanity. We have to bring our innocent,
sound, soulful side back to life in ourself and — in exactly the
same spirit — we have to bring innocent people to the fore in
our development. We have to acknowledge soundness now.
Those who can see most clearly should be out in front. An
automatic and inescapable consequence of the basic revela-
tion that our conscious mind was upset by the ignorance of
our conscience, is that some will have experienced more of
this upset than others, and that the least upset lead the way
home to a state of no upset. While the least upset lead the way
they will help the others and regulate their pace so everyone
keeps up (like Tiger did for Kweli) because no one is free
until we are all free.

At first we will strenuously resist acknowledging our
exhaustion and accepting leadership from the innocent.

Admitting and subordinating our upset goes against all our evasive power-and-glory-seeking old world practices. The truth is humans can't always achieve anything they want to achieve, being limited by our degree of alienation. In the old world we couldn't accept this truth because it implied we were bad and led to mental depression and to prejudice against the exhausted. Now we can accept it because we know it doesn't mean we are bad. We can be honest now. It is honesty that will free the world from upset.

Because the more exhausted, with more upset to be exposed, will resist the truth the most, the new world has to be brought to light by the less exhausted and by the young, who have yet to adopt evasion. Then they have to help the exhausted overcome their fear of the truth by showing them the love that this understanding brings. At present it is, in the main, the exhausted who are talking about the new world. They call it the New Age but the truth is the honesty required to bring it about means they will be the last to face the truth of our exhaustions and enter the new age.

> **Everyone who does evil** [becomes upset] **hates the light, and will not come into the light for fear that his deeds will be exposed.** [We could become so exhausted we could reach a situation where we preferred to stay in the dark; where we] **loved darkness instead of light.** [We could become] **a slave to sin.**
> *The Bible,* John 3:20,3:19,8:34.

Initially, many of us will see recognition of innocence as just a new form of prejudice — of elitism — of put-down of the less fortunate. After thinking about it however, we'll realise it's an essential and quite acceptable arrangement. Our freedom from upset depends on abandoning evasion, and now there is every reason to abandon it. Prejudice against upset and innocence has been removed; we differ in our degree of upset but we are all good.

The essential difference between the old world and the new is that in the new world we admit and subordinate our alienation. It is the honesty that is now possible — and nothing else — that can clear up and finally eliminate the suffering, the wars, the anger, the greed, the destruction of the environment and all the other manifestations of our upset and save the world.

Now that we know exhaustion or upset is not bad, we don't have to be ashamed of it in ourselves or critical of it in others. We can all see now that exhaustion developed from fighting for humanity. Our loss of innocence will now attract the heroic recognition it deserves, not the criticism it has for so long unjustly received.

We needn't fear leadership by the innocent: **If anyone wants to be first, he must be the very last, and the servant of all.** (*The Bible*, Mark 9:35.) The innocent are the least likely to want to wear crowns or to oppress or to exploit others. They will be the 'servant of all'. This is why they are the true leaders now. They know the way.

The innocent will grapple with the truth at close quarters, but won't hold that against the exhausted who clean up the environment. The exhausted won't envy, or feel any need to compete with the innocent. Everyone will know that their contribution to establishing the new world is as important as anyone else's. We will all be together in our different degrees of exhaustion. Understanding that we are equally good and being able to see what we have been through will bring great compassion and fellowship to our species. Fellowship is the answer to the problem of exposure. Adjusting to the new world is not only possible, it will be the most wonderful experience many of us have ever had. It simply requires a little patience for us to see and accept how it works.

The six stages most people — and in fact humanity as a whole — will go through in adjusting to the truth are:

First: accepting the explanations, initially because they make so much sense, then through experiencing their truth.

Second: rediscovering depressions that these ideas lead our mind back to: Gerard Manley Hopkins's **O the mind, mind**

has mountains; cliffs of fall Frightful, sheer, no-man fathomed.
(From his sonnet *No worst, there is none*, 1885.) A typical response
at this stage is 'Gee, I can see now why we hid from these
ideas'.

Third: rejecting the explanations, trying to avoid those
depressions. It is during this procrastination stage that the
symbol of the Foundation For Humanity's Adulthood — the
key held aloft — becomes vitally important. Those who can
must support the explanations even though others try to
ignore, deny and even attack them.

Fourth: realising that our denial is perpetuating suffering
on Earth and standing between humanity and its freedom,
and that denying the truth is not necessary now that upset is
defended.

Fifth: choosing one of the ways we can support the truth
without confronting it, that allows us to avoid depression
without denying the truth. We decide to let the truth through
past our upsets that want to deny it. This is our moment of
conversion to the new world.

Sixth: discovering the optimism, enthusiasm and fellowship that comes once these adjustments to the new world
have been made.

Essentially, the shock of being confronted with the truth
wears off and then we benefit enormously from being able to
understand ourselves and be honest about ourselves at last.
Being honest is the opposite of being alienated. We never
enjoyed having to be false, having to live in mystery and
confusion. Procrastination is a passing stage. Relief, excitement, fellowship and happiness is the real outcome of learning the truth about ourselves. Nevertheless these initial adjustments will not be easy and everyone making them should
have supportive counselling which has already proved valuable to many people who are in turn now counselling others.
Without supportive counselling it's all too easy to misinter-
pret the many new unevasive understandings, become slightly
lost again and start seeing ourselves as worthless again, which

could lead us back into depression. If we feel worthless we have necessarily misunderstood because <u>we are in no way bad or worthless</u>. To deal with a more superficial example of the misunderstanding that is possible: no, we don't now revert to hunter-gatherers and we don't 'throw the baby out with the bath water' and reject all our technological advances.

There is a lot to explain and understand about our new world. The Foundation For Humanity's Adulthood is urgently doing everything possible to provide support and help. Videos showing people having their questions, concerns and predicaments explained unevasively are proving especially useful. It is as if we have to go back to school; a whole new unevasive interpretation of ourselves and our world is now available. It is no exaggeration to say that our old world libraries of evasively presented understandings have become museums overnight. The evasion in everything from the old world has to be exposed and altered to fit the unevasive new world.

There is a lot of work to do but most importantly a great fellowship will soon appear on Earth. Every face will be a face of support and comfort. This is why the people shown symbolising the four activities of the new world on page 168 are touching each other. Rather than putting their arms around each other in an old world show of affection, they're together almost unconsciously as all humanity lived before.

I repeat these lines because as well as being my favourite quote lines used in this book, they reveal much about what freedom from upset was once like and will be like again.

Matata and Kanzi are housed in a large six-room indoor-outdoor enclosure. Even though Matata was wild-born, both she and Kanzi continually seek out and appear to enjoy and depend upon human companionship. Consequently, human teachers and caretakers are with them throughout the day. Mild distress is evidenced by Matata, even though she is an adult, at the departure of human teachers with whom she has formed close relationships. She seeks to maintain close proximity to these

teachers as they work with her, <u>often simply sitting with an arm or leg draped across them. When they stay until evening she pulls them into the large nest that she builds and goes to sleep next to them.</u>

Note again that while we will come to specialise in one of the four activities of the new world, we will all participate in all the activities to some degree. Even the exceptionally sound will find they can only confront the truth for a certain time each day before needing to do something less confronting (such as posting some books or videos about the understandings to friends) and after that something non-confronting like meditating, helping repair the environment or helping with relief work for the starving in third-world countries.

Apart from the four activities that introduce the new world, another concern is for all those left impoverished and enervated by the egocentric greed and falseness of the old world. We must now all help those who can't begin to think or to contribute towards the moratorium because they are starving or otherwise physically incapacitated.

The four ways of introducing the new world will become the new industries. They will replace the old world's escapist, consumerist, artificially glorifying and egotistical practices. (The truth is, despite all our old world ways of trying to be ideal by containing and concealing our upset, everything we touched inevitably became an expression of our upset. If we walk along any street and look at the houses we see that this is true. For our world to change we have to change. Better forms of management of our upset was not a real solution.) Gradually as our upset subsides so will our old-world practices. The emphasis will swing towards more selfless pursuits.

On the table in front of me is a silver teaspoon with an ornately engraved handle. It is very much an old world teaspoon. The bright silver and the embellishment glorified us when the world unjustly condemned us. It 'said' we were wonderful when the world in its ignorance wouldn't.

Without such materialistic reinforcement we could not have sustained our effort to find understanding. Materialism wasn't bad, in fact it was most necessary, but now it will gradually become unnecessary. The time and money spent digging up the silver and embellishing the spoon can now be spent helping others. We deserved to be glorified but the time and energy spent seeking glory impoverished others. The human condition made us self-preoccupied or selfish. We can now look at that teaspoon and recognise that it is a two or even three starving Ethiopians extravagance.

Everything about us from the old world is saturated with our greed, with our mind's hunger for relief from criticism. Our egocentric or embattled conscious thinking self can now be satisfied with understanding of its goodness. We no longer have to prove we're not bad by meeting challenges and by creating and winning competitions. Our egos have been satisfied at the most fundamental level. The source 'dragon' of all dragons is slain. The 'devil' who/which, when we were innocent, was the coercion from the upset world to compromise our ideals, or, when we were exhausted, was the false implication from the ideals that we were bad, is overcome.

Until now we have had to assert our ego or individuality. We have had to try to find understanding of our goodness and when we failed it was inevitable that we would try to satisfy our egos artificially through power, glory, fame, wealth and physical luxury. With our ego satisfied our needs aren't great.

We haven't been able to access the real enjoyment that our world has to offer, which is its great beauty and integrativeness, but now we can. In the not too distant future instead of making embellished silver spoons we will each carve from a piece of wood and carry our own simple spoon. One person can show another how to experience the beauty of sunlight and share the other's home-grown vegetables. They can then help build their community's simple communal mud-brick dwelling amongst the trees. The truth is cities were not functional centres as we evasively claimed, they were hide-outs for alienation and places that perpetuated/bred alienation. We will begin to close our cities down now.

> [Cain was the one who became upset/alienated and sure enough] **Cain was then building a city . . .**
> *The Bible*, Genesis 4:17.

> **The bush** [wilderness] **is our source of innocence; the town is where the devil prowls around.**
> Australian historian Manning Clark, *Sydney Morning Herald*, February 18, 1985.

> **'Do you see all these great buildings?' replied Jesus. 'Not one stone here will be left on another; everyone will be thrown down.'**
> *The Bible*, Matt. 24:2; Mark 13:2; Luke 19:44,21:6.

We have to rediscover the ability to feel. In *The Lost World of the Kalahari* Laurens van der Post says of the Bushmen, **He and they all participated so deeply of one another's being that the experience could almost be called mystical. For instance, he seemed to *know* what it actually felt like to be an elephant . . . a lizard . . . a baobab tree . . .** As has been mentioned, while the Bushmen are relatively innocent they are still members of modern, highly embattled Sophisticatedman, so if we are capable of this sensitivity while living naturally (in the environment our soul is familiar with) today, how much more sensitive must Childman, who had not even engaged in the battle, have been! Their sensitivity would have been so great it might appear to us as supernatural.

> I might say that many people who are born again to the soul's world are awestruck by the magnificence of this other world within us that they discover. Unfamiliar with it, they *do* see it as supernatural when in fact it's ultranatural. It's not paranormal, something aside from the normal but the most normal of places. Our current upset state is the abnormal, distorted one. Superstition and talk of our soul's world as being some remote realm of mystical spirituality is what I call 'exhaustionspeak'.

Sometimes when people became extremely exhausted their alienation (mental blocks) became disorganised and through this 'shattered defence' the soul occasionally emerged. They became 'mediums' or 'psychics' or 'channellers' or 'visionaries' or 'oracles' or 'people possessed' or 'people experiencing divine visitations' but it is a weird, disjointed expression of the truth that comes from access to the soul through a shattered defence. Christ for example was innocent not exhausted and he was not superstitious, nor did his knowledge of our soul's world appear in any way weird. The way the exhausted talk about our soul's world is very different from the way an innocent talks about it. To an innocent it is a natural place. Their love of this world or enthusiasm (the word enthusiasm comes from the Greek *enthios* which means 'God within') is natural. We have not differentiated between true prophets and false prophets who dabble in mysticism, fortune telling, incense-burning and the occult.

The ritual and reverence the exhausted surrounded the soul with hid its naturalness.

In 2 million years of battling and repressing our soul we have become extremely superficial and insensitive, almost completely numb beings.

As the causal memory [consciousness] **becomes more perfect in the primates, so the senses degenerate . . .**

In man sense-degeneration has reached an extreme point. Hypnosis proves, however, that this degeneration in man is not organic, or even functional in the generally accepted sense of the term. The organs are still capable of a very high degree of sensitiveness, and under hypnosis they may actually become functional. This sensitiveness must, therefore, be inhibited by the high mentality, and when this mentality becomes dormant under hypnosis the inhibition [alienation] **is removed . . .**

Subjects: *M.B.*, **a 21-year-old Boer girl; fairly well educated; . . . neurotic temperament.** *Several male and female chacmas* [baboons].

Twenty different people, the majority unknown to the girl, each handled a different small object and then placed it in a

receptacle. The girl, blindfolded, took out one object after another and by smelling the object and the hands of the different people handed each object back to the person who had handled it first without a mistake . . . A blindfolded chacma could not recognise an acquaintance standing within a yard. . . .

A sound of constant volume imitating the hiss of a snake, and in the case of the chacmas associated with the presence of an imitation rubber snake, could be clearly heard by the hypnotised girl at a distance of 230 yards. The distance at which average normal human beings could hear the sound lay between 20 and 30 yards. The chacmas could hear it at a distance of between 50 and 65 yards.

In distant vision the superiority of the chacma seemed greater still. A young captive male could at a distance of six miles, over a landscape flickering with mirage, recognise without fail among a group of people a human friend to whom he was greatly attached. At that distance in such atmospheric conditions no normal human, even with a good pair of binoculars, could distinguish human beings from cattle. . . . The hypnotised girl even at half that distance failed to recognise an acquaintance.

[Apparently humans had a superior sense of smell and hearing but inferior distance vision to baboons. Possibly our infancy and childhood wasn't spent in open country where baboons live, otherwise we would surely have had selected distance vision equal to theirs.]

George McCall Theal mentions the wonderful homing instinct of the Bushmen. Young children taken by wagon great distances from their homes found their way back through pathless wildernesses. This same 'instinct' is present in most primitive [innocent] peoples . . .
Eugène Marais, *The Soul of the Ape*, written in the 1930s, published in 1969, from chapters 7, 7 and 3 respectively.

Eugène Marais has been quoted before on page 34. On the back cover of my copy of *Soul of the Ape*, Robert Ardrey describes Marais thus: As a scientist he was unique, supreme in his time, yet a

worker in a science yet unborn. With the discovery of the reason for upset, unevasive science — of which Marais was an exceptional exponent — has now been born.

We now repair and then savour the world. Two million years of battling ignorance can subside. We can rediscover the true world of immense beauty and sensitivity that we lost access to.

For the first time in the history of consciousness we can abandon the battlefield. It is no longer a sign of weakness as it was when the battle still had to be won. For the first time, transcending our upset and being born again to the ideals are acts of total strength. Soon anyone playing egotistical, competitive, self-glorifying or selfish old world games will be pitied like people wearing last season's fashions. This may seem to imply elitist pressure but it is concern, not criticism, that will come through now. This book's essential vision is that soon an army of millions will appear to do battle with human suffering and its weaponry will be explanation, which is understanding.

The truth, which we will soon come to appreciate, is that being able to abandon our old world egocentric, competitive, self-glorifying lifestyle will be a great relief and not something difficult to do. Serving others is the most natural and satisfying attitude of all. In an Israeli kibbutz, all of which are 'new world' in some aspects of their structure, the street-sweeper and the Nobel Prize winner receive the same wage. The reward is the satisfaction of serving the common good (larger whole), which in the new world will be humanity.

Two great rewards awaiting each of us now are relief at being able to serve at last and feeling our upset subside as understanding gradually untangles our confusion and insecurity. In the four ways of participating formulation, we will also be gradually confronting our condition as well as transcending it, which will begin eliminating our upset. After about a year of this we will be able to realise how much we have changed, how much peace and relief has come to our

184

mind. If asked then what our situation would have been without these understandings we will say 'I hate to think'. We will perceive our great progress in eliminating our anger, egocentricity and alienation. Those who know us will see the difference and be impressed. The young will derive such confidence and certainty from understanding the world that older people may initially think they are precocious. In the old world wisdom (awareness that no one is bad) only came the hard way, after many years of experiencing the human condition. Now, nine-year-olds can be shown how to understand the human condition!

Paradoxically, in the final days of our struggle, religious and socialist principles of living by, and serving, the ideals, lost favour. Self-fulfilling free-enterprise and completely self-deluding attitudes swept the world. Now, principles of serving and honesty return.

There is no longer a left and right wing in politics, only a new form of left wing, completely free of pseudo idealism with its denial of exhaustion. (I stress that these changes won't be forced. They will happen naturally as we digest and share our new-found understandings. Revolution is not necessary. All we need do to overcome resistance and suffering is spread the understanding. It will create, not impose, the changes we seek, in ourselves, in others and in the world, and alleviate the world's distress.)

While this new way of living is not another religion — there is no deity worship or guilt in any form — it is like a new Church, but one that works fully. In the new world everyone will support the ideals every day while we gradually confront and digest the understandings of ourselves that are now available.

The important difference with this new 'Church' is that instead of supporting the embodiment of the ideals such as Christ, Buddha, Lao Tzu, Mohammed or the great Hindu prophets, we now support the understanding of the ideals, which brings us close to being the ideals. To use the terms of the exceptionally unevasive Pierre Teilhard de Chardin, we now move from the **Biosphere** to the **Noosphere** (Mindsphere) which is but a short step from becoming the **Body of Christ** or completely sound in self.

Humans were given consciousness without understanding, a computer with no program. Now that we have found the program all information will start to make sense. Soon we will be free of insecurity, evasion and falseness. Since our various personalities are in the main our various states of upset, in the upset-free future we will all have similar personalities.

Our mind or spirit has found understanding. All that remains is to absorb it.

> **. . . another Counsellor to be with you forever — the Spirit of truth . . . will teach you all things and will remind you of everything I have said to you.**
> *The Bible,* John 14:16,17,26.

> **Though I have been speaking figuratively, a time is coming when I will no longer use this kind of language but will tell you plainly about my Father.**
> *The Bible,* John 16:25.

> **. . . you will be like God, knowing . . .**
> *The Bible,* Genesis 3:5.

> **In the future they will every one be Buddhas.**
> **And will reach Perfect Enlightenment.**
> **In domains in all directions**
> **Each will have the same title.**
> **Simultaneously on wisdom-thrones**
> **They will prove the Supreme Wisdom.**
> Buddha (Siddartha Gautama) 560–480 BC, *The Lotus Sutra,* from Chapter 9. Translation by W.E. Soothill.

It is clear from the above that both Christ and Buddha looked forward to a time when understanding would replace dogma. The explanations in these pages confirm the words of the prophets; <u>religions aren't being threatened, they are being fulfilled</u>. I should add that our beliefs in an after-life and in a comforting God aren't destroyed by these scientific interpretations, which at first may appear cold and clinical. We do live on after we die although not literally as many believe. While we do live on in all things we don't come back in the body of a particular person or creature. Vincent van Gogh described the immortality of the human spirit succinctly when he wrote to his brother Theo that **while men live, all men live** (written between 1872–90, and mentioned in the film *Vincent*, 1987). In time we will come to see that our concepts of God and after-life haven't been destroyed, but confirmed, reinforced and enhanced. The difficulty is we are so extremely exhausted it's almost impossible for us to hold onto anything other than highly simplistic expressions of the truth.

Now that the largely male-dominated battle to champion the intellect or ego is won we can return to a matriarchal (female-role dominated) world where nurturing or love is all important.

The souls of little children are marvellously delicate and tender things, and keep forever the shadow that first falls on them . . . The first six years of our life make us; all that is added later is veneer . . .
Olive Schreiner, *The Story of an African Farm*, 1883.

As a result of accepting the significance of nurturing in our lives parenting will take on new meaning, responsibility and importance. This awareness, together with our emerging freedom from upset and resulting ability to care, will lead to restraint in our rate of reproduction.

We can now solve drastic problems like the burgeoning world population because we can address them at their source psychological level.

Feminism anticipated the new ego-free, women-liberated world but could have little general impact until the ego was satisfied. The battle to overthrow ignorance and champion the intellect had to be won first. Until it was won the repression of innocence had to continue. Now that the battle is won the Feminist or Women's Liberation Movement, like the New Age and Environment movements is validated.

> **. . . She is content, confident and unresentful because she is also the love that endureth and beareth all things even beyond faith and hope. She knows that, in the end, the child will grow and all shall be well.**
> Laurens van der Post, *Jung and the Story of Our Time*, 1976.

What was missing from the New Age, Environment and Women's Liberation movements was the means to self-confront. For our soul and the sensitivity and beauty of its world to be liberated we first had to find the understanding of our greater goodness. Our soul, in all its forms, has not been attacked and repressed without reason.

What was missing from the new world's infrastructure was the means to self-confront; its psychological base. The key to the new world is the explanation of our upset. We had to solve the riddle of human nature; discover why we were so divisive when the ideals are clearly to be integrative. This understanding was the Holy Grail we sought as a species. With it found, the new world can at last be born properly. The impasse to the new world is now breached. The floodgates are open. The new world now comes in a rush.

> **. . . your kingdom come,**
> **your will be done**
> **on earth . . .**
> *The Bible*, The Lord's Prayer, Matt. 6:10.

The ability to be unevasive, to be honest, to confront the truth sets us free. The truth comes not to hurt us but to free us. Honesty is therapy . . . **the truth will set you free** (*The Bible*, John 8:32), but it had to be the full truth that defends us.

For copyright reasons I am unable to reproduce the lyrics of John Lennon's song *Imagine* (1970), paraphrased in part as follows:

Imagine everyone sharing everything. Even though you might think I'm dreaming I'm not alone. I hope you'll join us and we'll all live as one.

... [we will] **change and become like little children** ...
The Bible, Matt. 18:3.

The explanations presented in this book are dealt with in greater depth in *Free: The End of The Human Condition* which, as a result of feedback from readers and elaborations, is due to appear in an Expanded Second Edition in 1992.

For:
- copies of *Free: The End of The Human Condition,*
- copies of the Expanded Second Edition when they become available,
- copies of *Beyond The Human Condition,*
- promotional material such as stickers and Adam Stork T-shirts,
- a subscription to the Foundation For Humanity's Adulthood. In 1991 the annual fee is $A30, which brings you some six newsletters a year, information on the Foundation's courses and seminars, access to the library, and video and audio tapes that help explain and introduce the information):

<u>Telephone</u> (02)94863308, Fax (02)94863409. If calling from outside Australia, dial your country's international access code then 61-2-94863308 or, for fax 61-2-94863409.

<u>Write</u> to the Foundation For Humanity's Adulthood, GPO Box 5095, Sydney NSW 2001, AUSTRALIA. (The box number ensures the Foundation a permanent address should its location and phone number change.)

Website: **www.humancondition.info**

The Foundation For Humanity's Adulthood

The Foundation is a non-profit organisation established to promote the explanations given in this book.

In a world in crisis the need for this dignifying understanding of our species is clear yet the profoundness of the ideas makes even their consideration difficult. The task before us is immense.

Our immediate objectives are to attract scientific interest in the ideas, publish *Beyond The Human Condition* internationally, produce explanatory videos and a television documentary, develop study courses and establish wilderness schools (the natural environment is the ideal setting in which to introduce these concepts). In addition elaborations of these ideas need to be edited and published, suscribers and supporters catered for and our counselling services expanded.

Currently the Foundation is funded by subscriptions and book sales, but further funding is needed. We are especially looking to the strong, rational business community for support.

Taking humanity beyond the human condition will require the help of everyone. If you are able to assist in this vital pioneering stage please contact the Foundation.

Tim Macartney-Snape
(Director).

Index